D1082482

A CULTURAL HISTORY OF MODERN SCIENCE IN CHINA

NEW HISTORIES OF SCIENCE, TECHNOLOGY, AND MEDICINE

SERIES EDITORS
Margaret C. Jacob, Spencer R. Weart, and Harold J. Cook

BENJAMIN A. ELMAN

A CULTURAL HISTORY OF MODERN SCIENCE IN CHINA

HARVARD UNIVERSITY PRESS

CAMBRIDGE, MASSACHUSETTS

LONDON, ENGLAND

Printed in the United States of America

First Harvard University Press paperback edition, 2008

Library of Congress Cataloging-in-Publication Data

Elman, Benjamin A., 1946–
A cultural history of modern science in China / Benjamin A. Elman.
p. cm.—(New histories of science, technology, and medicine)
Includes bibliographical references
ISBN-13: 978-0-674-02306-2 (cloth: alk. paper)
ISBN-10: 0-674-02306-4 (cloth: alk. paper)
ISBN-13: 978-0-674-03042-8 (pbk.)
ISBN-10: 0-674-03042-7 (pbk.)
1. Science—China—History. 2. Technology—China—History.
I. Title. II. Series.

Q127.C5E46 2006
509.51′0903—dc22 2006042706

FOR MY STUDENTS,
AND FOR THOSE WHO WILL GO FURTHER

CONTENTS

CONTENTS

ILLUSTRATIONS

Maps

ILLUSTRATIONS

Figures

CONVENTIONS

Like *On Their Own Terms: Science in China, 1550–1900* (Harvard University Press, 2005), this shorter history is not a topically organized historical survey of the Chinese sciences; Joseph Needham's *Science and Civilisation in China* series is the best sourcebook for that approach. Instead, in this book I focus principally on Chinese natural studies and the literati mastery of European natural learning from 1600 to 1900. In particular, I show how Manchu rulers and Chinese scholars broadened late imperial studies of astronomy, geography, mathematics, and medicine through court and literati contacts with Jesuits and Protestants, who presented European natural philosophy as part of their religious inheritance.

Chinese characters have not been included for Chinese or Japanese historical figures. Most are readily available in biographical collections such as the *Dictionary of Ming Biography*, ed. L. C. Goodrich et al., 2 vols. (New York: Columbia University Press, 1976); *Eminent Chinese of the Ch'ing Period*, ed. Arthur Hummel (reprint; Taipei: Chengwen Bookstore, 1972); and Howard Boorman and Richard Howard, *Biographical Dictionary of Republican China*, 5 vols. (New York: Columbia University Press, 1967–1971). A bibliography of Chinese and Japanese sources is presented in *On Their Own Terms*. See also the Abbreviations to the Notes and Appendixes in this volume.

Terminology such as "the West" refers to early modern Europe (1600–1800) in Chapters 1–3 and grows to include the United States and modern Europe (1800–1900) in Chapters 4–7. Chinese sources typically refer to the "West" or "Western." When it is appropriate, I narrow the focus to Western European in Chapters 1–3. Normally, I present the "West" as a cultural construct used by Jesuits, Protestants, and Chinese to refer to themselves or to the "other." "European" and "Euro-American" will designate a regionalized view with more historical and chronological precision.

The role of the *Change Classic* (*Yijing,* or "Book of Changes," also known as *I Ching*) in Chinese natural thought and science will not be stressed in this volume. Its cosmological and metaphysical influence on Chinese conceptions of natural phenomena through the late imperial period was pervasive. But its effects on those Chinese who increasingly turned to modern science waned. This group of forward-looking scholars was always a minority in the larger sea of Chinese who adored the multifaceted and open-ended interpretations of the *Yijing,* which explained and predicted the corresponding changes in the social, political, and bodily realms. But this relatively small minority in 1800 became the cutting edge in 1900 that rightly or wrongly tipped the balance and implicated the *Yijing* in the premodern cast of Chinese traditional science. Interested readers can refer to the important work of Richard Smith ("The Languages of the *Yijing* and the Representation of Reality," *Oracle* [Summer 1998]: 35–50) and Benjamin Wai-ming Ng ("The *I Ching* in the Adaptation of Western Science in Tokugawa Japan," *Chinese Science* 15 [1998]: 94–117) for this aspect of traditional natural studies.

CHINESE DYNASTIES

Shang: 16th–11th centuries B.C.E.

Zhou: 11th century–221 B.C.E.

Qin: 221–207 B.C.E.

Han: 206 B.C.E.–220 C.E.
- Former Han: 206 B.C.E.–8 C.E.
- Later Han: 25–220

Wei: 220–265

Western Jin: 265–316

Southern and Northern Dynasties: 386/420–581
- Liu-Song (Southern Dynasty): 420–479
- Northern Wei: 386–534
- Northern Zhou: 557–581

Sui: 581–618

Tang: 618–907

Liao: 916–1125

Song: 960–1280
- Northern Song: 960–1127
- Southern Song: 1127–1280

Jin: 1115–1234

Yuan: 1280–1368

Ming: 1368–1644

Qing: 1644–1911

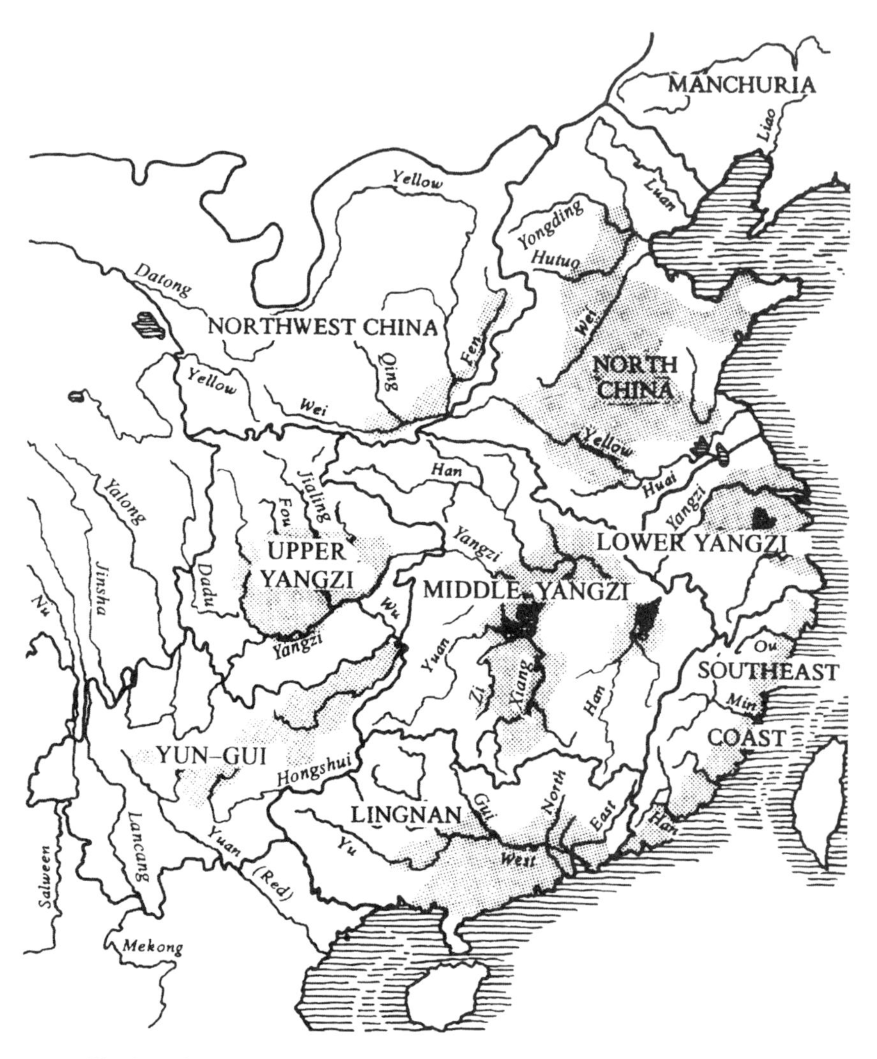

Physiocratic macroregions of agrarian China in relation to major rivers. From G. William Skinner, ed., *The City in Late Imperial China.* Copyright © 1977 by the Board of Trustees of the Leland Stanford University.

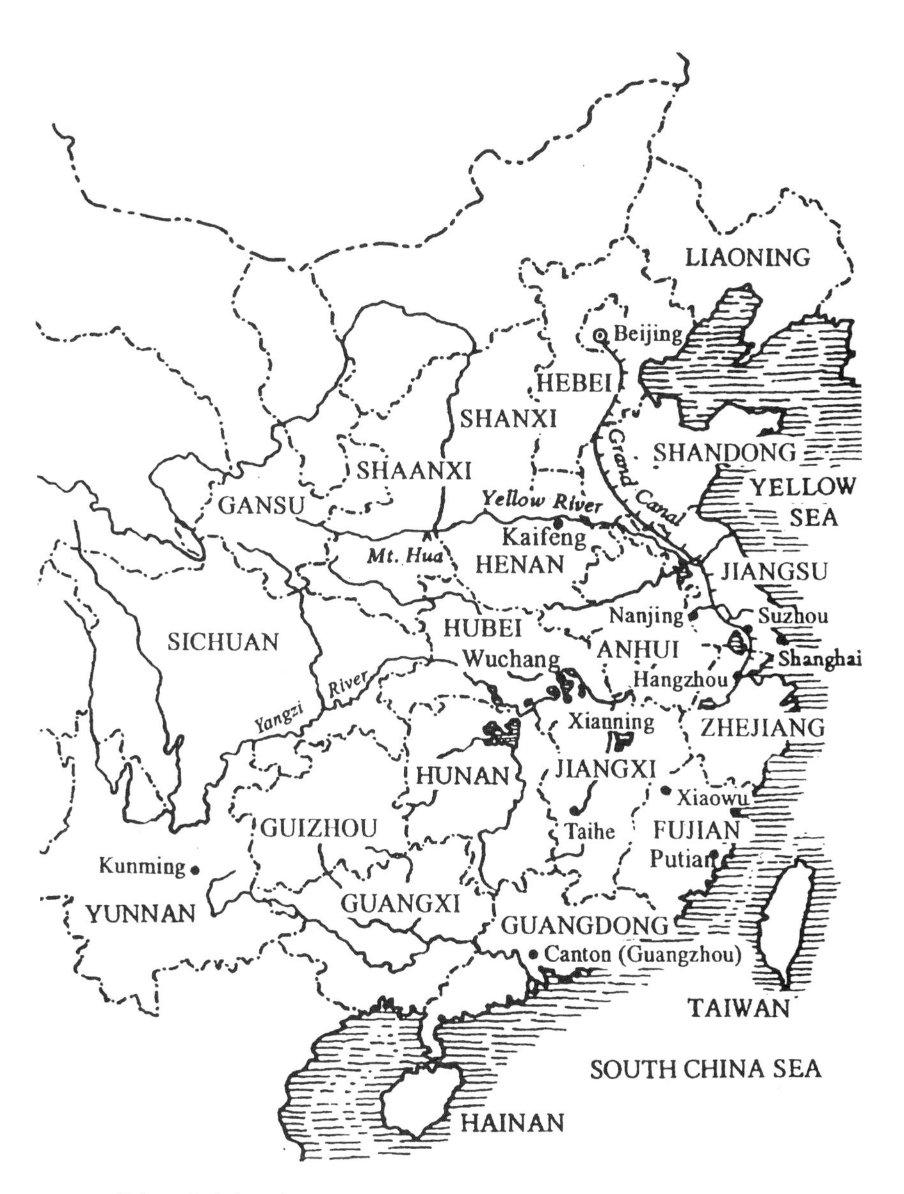

China administrative map.

INTRODUCTION

This textbook describes the cultural history of modern science, medicine, and technology in China since 1550. It ties together the Jesuit influence on late imperial China, circa 1600–1800, and the Protestant era in early modern China from the 1840s to 1900. If there has been one constant in China since the middle of the nineteenth century, it is that imperial reformers, early Republicans, Nationalist party cadres, and Chinese Communists have all made science and technology a top priority. Yet we have undervalued the place of science in modern Chinese history. Indeed, our understanding of all Chinese natural studies before 1900 has remained the domain of specialists.

Overview

Chapter 1 explains that many Chinese literati were extremely curious about European science in the seventeenth and eighteenth centuries. Jesuits in China since the 1580s had accommodated differences between European and Chinese learning by focusing on mathematics and astronomy, an approach that they generally did not employ in Japan, India, Persia, or Southeast Asia, not to mention the New World. To gain the trust of the imperial throne and its literati elites, Matteo Ricci (1552–1610) and his colleagues prioritized natural studies and mathematical astronomy during the late Ming (1368–1644) and early Qing (1644–1911) dynasties precisely because they recognized that literati and emperors were interested in such fields.[1]

The literati in particular represented select members of the landholding gentry who maintained their status as cultural elites primarily through classical scholarship, knowledge of lineage ritual, and literary publications. As members of the gentry, they also wielded local power until 1900 as landlords or provincial leaders and empirewide power as government officials. The cultural status of both the gentry at large and the literati in their midst correlated with their rank on the civil service examinations. In addition, during the late empire, gentry and merchants intermingled, with merchants becoming part of the gentry elite. The Jesuits realized that by promoting natural studies they could improve the cultural environment for converting the Chinese to Christianity. This accommodation lasted until the Rites Controversy in the early eighteenth century, when Catholics in Paris and the Beijing court debated whether Chinese ancestor worship was fundamentally religious or civil in character, and whether Chinese terms for God taken from the literati canon were appropriate. Throughout the period of Jesuit influence, European enemies of the Jesuits accused them of making themselves useful to local rulers (such as the newly enthroned Manchus) for personal advantage rather than in the name of Christianity.

The failure of the Jesuit mission and other European endeavors to transmit scientific and mathematical knowledge during and after the Kangxi reign (1662–1722), when native literati challenged the Jesuit monopoly of early modern astronomy, was not due to lack of Chinese interest. Rather, the Chinese absence of further knowledge about eighteenth-century scientific developments in Europe resulted from a break in scientific transmission caused by the demise of the Jesuits worldwide and their schools in Europe during the eighteenth century—a demise that, for example, delayed for almost a century the relaying of information from Europe about the role of calculus for engineering and mechanics for physics.[2]

Historians of science have generally presented the Jesuit and Protestant missions to China as two separate events, leaving a gap in the history between 1780 and 1840. I fill in this gap by describing the Jesuit era in China, circa 1600–1800, as an important preface to the Protestant influence from the 1840s to the 1890s, when modern science took root.[3]

I will also show that in many cases Jesuit influence was not especially relevant to ongoing changes in literati learning. In the medical field, for example, few Chinese physicians before the nineteenth century took early modern European Galenic medicine and its theory of bodily humors seriously as a threat or complement to native remedies. The Jesuits, although they were not the first Catholic order to establish scholarly links with the Chinese empire, subsumed their idiosyncratic, Aristotelian version of natural studies within their premodern version of "Western learning," which united religion and *scientia* (specialized learning). They and their Chinese collaborators translated such studies as "learning based on the investigation of things and the extension of knowledge."[4]

By contrast, the Kangxi revival of interest in mathematics was closely tied to the introduction of Jesuit algebra, trigonometry, and logarithms. Jesuits in the China mission during the eighteenth century brought a range of technical skills, from surveying methods to cannon-making. They also introduced pulley systems, sundials, telescopes, water pumps, musical instruments, clocks, and other mechanical devices. In addition, emperors, their courts, and literati families welcomed Western goods and manufactures, which even made appearances in Chinese novels of the time such as *Dream of the Red Chamber*.[5]

Initially, the Chinese required a higher degree of astronomical expertise than the Jesuits could provide: the imperial astronomers wanted to predict eclipses on the basis of notions of cyclical time, whereas Gregorian reformers in the 1570s had not worried about eclipses to determine the repeating but ultimately linear date for Easter. Because of the Chinese preference for better prediction, after Ricci's death the Jesuits in China went beyond the Ptolemaic earth-centered (geocentric) world and mastered the new Tychonic (geo-heliocentric) world system of an earth-centered cosmos around which the sun and its planets revolved. Once this was accomplished, however, the work of Jesuits in the imperial Astrocalendrical Bureau became regulatory rather than explorative.

Armed with the intellectual and instrumental tools provided by Tycho Brahe (1546–1601) and his followers, the Jesuits solved the problems they were hired to undertake. But they did not keep up with newer scientific developments in Europe, which more and more were products of northern European Protestants outside the Church. Consequently, the Tychonic system was still used to train astronomers in Qing China during the late nineteenth century. The Kangxi emperor reproduced the institutional models for translation that Ming Chinese scholars such as Xu Guangqi (1562–1633) had created with the help of Matteo Ricci and Li Zhizao (1565–1630). After 1850, on an even larger scale, the Protestant missionaries collaborated with the Chinese to translate postindustrial revolution science and technology into Chinese.[6]

Chapter 2 assesses the growing interest among elites in classical medicine, mathematics, and astronomy during the eighteenth century. In 1814, when Pope Pius VII restored the Jesuits as a religious order after the defeat of Napoleon, it was in the context of the Congress of Vienna and its efforts to reinstate the old order across Europe. In this climate, the Jesuits became even more conservative and generally opposed the progressive-liberal ideologies of the Protestant missionary organizations sent to China after 1840.

The breakdown of the Jesuit consensus in the eighteenth century coincided with an increasing Chinese self-reliance in mathematical training. Literati contended that European learning was rooted in China's ancient classics. Under imperial patronage, literati upgraded mathematical studies from an insignificant skill in 1700 to an important domain of knowledge that by 1800 complemented classical learning. One irony of the failure of the French Jesuits to keep up in mathematics and science was that although French Jesuits had corresponded with Gottfried Leibniz (1646–1716), inventor of the notational forms for the calculus that engineers employed during the eighteenth century in Europe, the Jesuits failed to see beyond the Christian mysteries they read into the mathematical arrangement of the sixty-four hexagrams of the *Change Classic (Yijing),* a divination text of possible imminent transformations that became one of the Five Classics during the Han dynasty (206 B.C.E.–220 C.E.).

Chapter 3 recounts how Protestant missionaries and their Chinese assistants finally translated into Chinese Newton's laws of physics in the 1687 *Principia,* as well as modern mathematical analysis, when the missionaries arrived in China's newly opened treaty ports after the First Opium War (1839–1842). The leaders of the 1793 Macartney mission from Great Britain defined their historical role vis-à-vis their Jesuit predecessors by presenting—both to the Manchu court and to Chinese literati—Great Britain as the manufacturing leader of Europe and themselves as enthusiastic teachers of their new scientific knowledge.

Lord Macartney (1737–1806) believed that the gifts that he had brought, chief among them a solar planetarium synchronized by "the most ingenious mechanism that had ever been constructed in Europe," were more sophisticated than the heliocentric armillary spheres, mechanical clocks, and telescopes previously introduced by the Jesuits. He also thought that such gifts would convince the Qianlong emperor (r. 1736–1795) of Britain's dominance in science and technology.[7]

But Qing China did not yet require the specific "ingenious objects" or "manufactures" from England that Macartney presented, any more than India had needed such things before the Manchester textile factories exploited the Indian cottage clothing industry. Such material needs in China only developed after an illegal opium trade created a commodity that millions of Chinese could not do without.

Because of Britain's role in defeating Catholic France, British Protestant missionaries increasingly traveled to India, Singapore, and China. In the late eighteenth century, Protestant churches became influential arbiters of public attitudes toward social issues of the day. Evangelism, the belief in salvation by faith alone, was formed under the auspices of Protestant churches, particularly the Methodists and low-Church Anglicans, and informed Victorian public values. Meanwhile, Great Britain's scientific ethos, which was based on Newtonian mechanics and machine-driven industry, was transmitted worldwide by enlightened missionaries and Church evangelicals.[8]

A resurgence of emigration in the mid-eighteenth century paralleled growth in the British Empire. The middle classes joined in the nation's civilizing mission at home and abroad. Indeed, the 1780s and 1790s were a time when missionary philanthropy for spiritual conversion found widespread support, as the missionary enterprise turned from saving lives through helping the poor at home to reforming manners and saving souls abroad.

Protestant initiatives to introduce European and American political and cultural institutions to China in the 1830s built on and surpassed earlier Jesuit geographies of the world and descriptions of European countries. In turn, Chinese scholars based in Guangzhou gained a more precise understanding of the new missionaries and the various countries they came from. The Qing dynasty developed strategic information to deal with the opium menace, even though a maritime crisis was also at hand. The Chinese realized that their knowledge of the West, which had been acquired through interactions with the Jesuits during the seventeenth and eighteenth centuries, was obsolete.[9]

Chapters 4 and 5 explain how an emphasis on science for dynastic self-strengthening attracted the interest of the Chinese literati, particularly after the debacle of the Taiping Rebellion—a massive peasant uprising fomented by anti-Manchu and anti-literati leaders that traumatized most of South China from 1850 to 1864 and caused the death of some twenty to thirty million Chinese. The Chinese terminology for modern science in the nineteenth century became less and less compatible with the Ming-Qing traditions of natural studies. But connecting state power to scientific advancement required something of a shift in mindset. The linking of state power to science had not occurred to the Jesuits—who communicated (and tailored) many of the early Western scientific breakthroughs to the Chinese—because Christianity still took precedence. They continued to link science to the Church and its religious mission.

From 1850 to 1870, a core group of missionaries and Chinese co-workers in Guangzhou, Ningbo, Beijing, and Shanghai translated many works on astronomy, mathematics, medicine, as well as botany, geography, geology, mechanics, and navigation. Alongside the efforts of the arsenals and official schools, private initiatives popularized modern science *(gezhi xue)* in the treaty ports and among Qing officials and literati. During the second half of the nineteenth century China's modernization, which was initially perceived as Westernization *(Xihua)*, began.

Chapter 6 describes the new arsenals, shipyards, technical schools, and translation bureaus, which all contributed to China's early industrialization. The increased training in military technology and education in Western science that was available in Qing China (a decade before similar efforts in Meiji Japan) allowed Qing reformers to forge a union of scientific knowledge and experimental practice after 1865 among literati and artisans—first and most notably in Shanghai at the Jiangnan Arsenal, but also at the Fuzhou Navy Yard. Japan's Iwakura Embassy noted both industrial sites when it visited Shanghai during its worldwide fact-finding mission from 1871 to 1873.[10]

Given their comparable levels of science and technology in the 1890s, Qing China might have defeated Meiji Japan in their great war over Korea in 1894–1895. When we look carefully at the total picture of the so-called Self-Strengthening period from 1865 to 1895, the view that Qing China was irrevocably weak and backward in contrast to a powerful Europe and a rapidly industrializing Japan is unfair. This perspective is a relic of the effects of the Sino-Japanese war on international and domestic opinion.[11]

Changes in elite and popular opinion after 1895 forced the Chinese literati to reconsider their strategy of adapting Euro-American science and technology to traditional culture. Westernizing radicals such as Yan Fu concluded that the accommodation between Chinese ways and Western institutions, which had informed the Self-Strengthening Movement since the 1860s, had failed. The Sino-Japanese War altered the frame of reference for the post-1895 period for the new Chinese and Japanese intelligentsia. For Chinese students of modern science and medicine studying overseas during the late Qing, Japan replaced the Jesuits and Protestants as the messenger of science.

Chapter 7 traces the beginnings of the failure narrative for Chinese science, in exploring why historians have historically undervalued the period described in this book. After 1900, Chinese elites, particularly reformers and revolutionaries, increasingly demeaned their traditional sciences as incompatible with the universal findings of modern science. They disparaged Chinese natural studies and medicine, transferring their support instead to a set of revolutionary institutional changes that required modern science and medicine based on Western models (as mediated by Meiji Japan).[12]

After 1895, reformers no longer heralded as Western-style enterprises the arsenals, navy yards, and factories where the manufacture of armaments and ships had begun. Instead, they dismissed them as backward Qing dynasty military factories. Single-minded promotion of Western learning caused a cultural rupture in attitudes in China and Japan. Radicals who emulated Meiji Japan (1868–1911) separated the goal of understanding modern science from the task of preserving the Qing dynasty.[13]

Because civil examination credentials no longer confirmed elite status after 1905, everyone—not just those who had failed the test—turned to avenues of learning and careers outside officialdom. Many traveled abroad or to established Chinese treaty ports such as Shanghai to seek their fortunes as members of a new gentry-based Chinese intelligentsia that would spawn a new generation of modern Chinese intellectuals, scientists, doctors, and engineers. The failure of the 1898 educational reforms meant that China came to rely even more heavily on Japan to translate—and mediate—Western scientific developments. The new texts translated from Japanese made modern science available to a wider audience and raised the level of knowledge among students and teachers.

Chinese students educated abroad at Western universities such as Cornell University or sponsored by the Rockefeller Foundation after 1914 for medical study in the United States—as well as those trained in the sciences locally at higher-level missionary schools and the new universities—regarded modern science in light of the Japanese version of Western science, rather than in terms of traditional frameworks, which they believed to be inappropriate for universal, modern science.

After 1915, the teleology of a universal and progressive science first invented in Europe replaced the Chinese notion that Western natural studies had their origins in ancient China. The dismantling of the traditions of Chinese natural studies *(gezhi)* and natural history *(bowu)*, along with many other categories, which had linked the premodern sciences and medicine to classical learning from

1370 to 1905, climaxed during the New Culture Movement from 1915 to 1919. When their opposition to classical learning and its traditions of natural studies peaked, New Culture advocates helped replace the imperial tradition of natural studies and classical medicine with modern science and medical therapies.[14]

Some Historiographical Issues

Historians of science, medicine, and technology have portrayed the period from 1500 to 1800 mainly through European frames of reference, even when their narrative accounts stress comparative themes. Because the emergence of modern science in the industrializing portions of Western Europe represents their central story, they have not probed how the interactions since 1600 between Asia and early modern Europeans over the meaning and significance of natural studies evolved from the Asian perspective.[15]

Eurocentric portraits of the rise of modern science, while not monolithic or one-dimensional, usually represent variations of a historical teleology of Western European scientific "success," and, by comparison, non-Western "failure." Often the narrative in such accounts reproduces uncritically the story of the Protestant-based scientific revolution of the late-seventeenth or early eighteenth centuries, or revisits the narrative of the medieval, Catholic roots of modern science.

This book moves the issue of science during the Ming and Qing dynasties from the background to the foreground of modern Chinese intellectual history. Consequently, natural studies in imperial China are presented in a context that also explores Jesuit learning and early Protestant translations. The historical transition to modern science in China during the late nineteenth century and the beginnings of the modern Chinese industrial revolution are also important themes.

Europeans increasingly thought themselves scientifically and technologically superior to others after 1500, but neither the Chinese nor the Japanese agreed with this perspective until they observed the military effects of the Industrial Revolution on nineteenth-century battlefields in East Asia. In early modern Japan, for instance, aristocratic and merchant elites domesticated European learning using classical Chinese and Japanese frames of reference (despite the influence of Dutch Learning and modern European science and medicine in the southern Japanese Dutch enclave of Nagasaki in the seventeenth and eighteenth centuries).[16] Within a colonial context, British imperial power in India after 1700 dictated the terms of social, cultural, and political interaction between natives and Westerners. Like in Japan, however, natural studies in late imperial China until 1900 were part of a nativist imperial project to master and control Western views on what constituted legitimate knowledge.[17]

During the transition from Dutch Learning to "foreign learning" in the nineteenth century, the Dutch in Nagasaki harbor no longer monopolized the transmission of the Western sciences to Japan. Moreover, the 1798 and 1842 calendrical reforms in Japan were based on Chinese translations of Western astronomy that had been completed with the help of the Jesuits appointed to the Qing Astrocalendrical Bureau. The Tokugawa shogunate (1600–1857) had banned the Jesuits from its islands early in the seventeenth century. In the early nineteenth century, English, French, and German works of science replaced Dutch translations and broadened Japanese interest in foreign learning. We will see in Chapter 7 that Meiji Japan still depended on Chinese translations of the sciences during the Protestant century in China, 1840–1895.[18]

In South Asia, the British colonial regime set the agenda for natural studies by defining the body of Western (that is, British) knowledge that local collecting procedures would augment. Such knowledge was in turn ordered and classified according to the standards of authoritative British scientific practice. Colonial forms of knowledge were translated into reports, statistical records, histories, gazetteers, legal codes, and encyclopedias that induced elites in India to become part of Britain's project of political and cultural control. Such knowledge was channeled through the Western structures of science, and colonized natives acquired enough practical experience to understand how to acquire, study, and interpret natural knowledge.[19]

Gyan Prakash describes how the British regime actually staged science in India via museums, exhibitions, and governmental projects. Such stagings presented science as a universal sign of modernity and augmented colonial rule by educating native elites according to the Western forms of scientific knowledge and natural history that were evolving in the nineteenth century. Western educated elites portrayed modern science and technology as a preferred value system and useful technology that could enrich India's indigenous traditions.

But Prakash also describes how Indian elites themselves renegotiated the terms of their acceptance of the British regime of science and technology. They created a hybrid discourse of science and nation, which claimed the ancient Hindu *Vedas* as the roots of science and inscribed Indian classical philosophy and religion with modernist forms. The colonized in India thus were not simply passive recipients of modern science; instead, colonized Indian intellectuals appropriated science in their own way, refashioning their own traditions in light of the ideals of modern science rather than in ways that always pleased the intimidating British colonial power.[20]

The colonized Indians may have offered some resistance to British interpretations of modern science, but late imperial Chinese literati and their government, whether under Ming Chinese or Qing Manchu rulers, effectively contested European claims to scientific and religious superiority at every stage of their interaction from 1580 until 1900. One of the reasons we have detailed accounts of the conditions in Chinese prisons in the sixteenth and seventeenth centuries, for instance, is that aggressive religious proselytizers from the Augustinian, Dominican, Franciscan, and Jesuit orders prepared the accounts after the Ming imperial state locked them up for interfering too much in state affairs and Chinese culture.[21] The native vicissitudes in the emergence of modern science in China, as in Japan and India, deserve a more nuanced account than is usually given in histories of the rise of modern science in Europe circa 1500–1800.[22]

Until the late nineteenth century, the Chinese and Manchus did not engage in the covert colonialist renegotiations with Europeans in which Hindu nativists in India indulged. At the same time, the Ming and Qing imperial court induced Jesuit mathematical, astronomical, military, and mensuration experts to work as minions in the government bureaucracy to augment each dynasty's own project of political and cultural control. Consequently, it would be a mistake to underestimate Chinese efforts to master on their own terms the Western learning of the Jesuits in the sixteenth, seventeenth, and eighteenth centuries.

Western modernization narratives have described British imperial expansion colliding with a sinocentric Qing state unsympathetic to scientific knowledge. This view should be amended. The famous 1793 letter by the Qianlong emperor (r. 1736–1795) to George III gainsaying Western gadgets should not be read as the statement of a Manchu empire out of touch with reality. The emperor did not categorically reject Western technology; in fact, Qianlong showed great interest in, for example, the model warship equipped with cannons that Macartney presented.[23] Rather, the mutual misunderstandings swelled from the overstated claims that Macartney made about his gadgets—which included a replica of the solar system. Now that the Qing calendar functioned properly with Jesuit help, the emperor was not inclined to think the planetarium so fabulous.

Later emperors who found English military firepower irresistible in the aftermath of the First Opium War were dealing with a different set of technological—and cultural—circumstances. Literati scholars had incorporated mathematical study and made natural studies a part of classical studies, but the fate of natural studies and technology in China after the Jesuits was highly influenced by the modern sciences introduced by Protestant missionaries, which included ideas for improving China's military defenses from post-Napoleonic Europe.[24]

We do not have appropriate categories of learning that resemble the premodern Chinese frames for what we call natural studies or natural history, and according to which Chinese literati evaluated Jesuit and Protestant "Western learning" and science during the Ming and Qing dynasties. An analytical ordering of early modern European scholarship, such as the Jesuit *scientia,* within the framework of modern learning is equally problematic and anachronistic.[25] Thus to understand premodern Chinese knowledge systems of the natural world, or for that matter the mindset of scholars in early modern Europe, requires that we try to appreciate these sets of ideas and beliefs as plausible for those living at the time. Placing the field of natural studies in China within its own internal and external contexts enables us to reconstruct its historical communities of interpretation and consider how those communities later recast modern science in Chinese terms.[26]

Beneath the cultural narrative of scientific, technological, and military failure, which many Euro-Americans, Japanese, and Chinese still attach to Chinese history after 1895, lies another story. It will never replace the triumphal recounting of the march of Western imperialism via science, technology, and empire-building. Nevertheless, this quieter story is important for illuminating longstanding Chinese interests in the natural world, the arts and crafts, and commerce that set the stage for interactions with European science, technology, and medicine until 1900.

THE JESUIT LEGACY

When Jesuits from Europe arrived in Ming China, Chinese literati were debating their own traditional theories of knowledge. Some claimed that morality took precedence over formal knowledge. Others contended that formal knowledge prevailed. Each side focused on the "investigation of things and the extension of knowledge" as the grid for knowledge and morality. Late Ming scholars tried to reconstruct the textual lives of things, phenomena, and events at a time of increasing commercialization known as the silver age. The human interest in natural and manmade things had led to commodities for the many, which some felt betrayed the ideals of moral cultivation.[1]

The Jesuits added precision to the investigation of things by exposing seventeenth-century literati to early modern European classifications, forms of argument, and organizational principles for specialized knowledge, that is, *scientia.* When the Chinese learned about Western learning in the late Ming, they equated it with ancient literati classical learning. In addition, they often stressed that study of the two traditions necessitated "exhaustively fathoming principles and investigating things."[2]

The wax and wane of Jesuit influence coincided with the sudden change in dynasty from Ming to Qing. Jesuits found that their literati counterparts resented their increased influence under early

Manchu rulers and their collaborations with well-organized bannermen armies of Manchus, Mongols, and northeastern Chinese. Astronomy became the focal point for the contending groups to shape their political influence. It was in mathematical astronomy, not medicine, that the early Jesuits in China would leave their mark, especially during the astronomy crisis that brought the Kangxi emperor to power.

Scholarly debates between the literati and the Jesuits intensified during the early Qing dynasty, particularly after 1660 when Christian creationism and biblical chronology challenged Chinese assertions of their cosmic centrality. In addition, fundamental differences—between the orthodox study of principles among literati and Jesuit views of how God had created the world and endowed humans with an eternal soul—became political flashpoints. In the 1660s, the Manchu court's brief tolerance of Jesuit religious beliefs was severely tested when Chinese, Muslims, and Manchus attacked Jesuits and their staff in the Astrocalendrical Bureau as enemies of both the dynasty and classical civilization.

Although the Jesuit mission survived the 1664–1667 purges, from then on the Manchu court carefully scrutinized its missionary impulses. As long as the Jesuits provided the court with needed expertise in mathematical astronomy and geographical matters, the emperor kept them on, but increasingly the needs of the Qing government, rather than the goals of the Church, defined their activities.[3]

The Late Ming Calendar Crisis

As the "Son of Heaven," the Ming emperor was responsible for an accurate calendar to order the empire. To accomplish this, the emperor relied on a precise astronomical system known as the Grand Concordance to understand the correspondences among the heavens, the earth, and humans. Chinese and non-Chinese dynasties employed court astronomers who could interpret events such as nonperiodic stellar novae, solar halos, meteors, and meteorites, as well as predict periodic lunar and solar eclipses. The imperial court had long appointed expert foreigners such as Persians and Muslims to the Astrocalendrical Bureau to disseminate an official lunisolar calendar. Celestial events that imperial officials could not predict, as well as earthquakes, famines, and so on, were portents that potentially pointed to the emperor's lack of virtue and his possible loss of the Mandate of Heaven.[4]

Both the dynasty in power and the people at large were also concerned with auspicious and inauspicious dates for enacting events, actions, and rituals. In 1592, however, the Ministry of Rites charged that the Astrocalendrical Bureau had erred by a full day when predicting a lunar eclipse, which meant that the ceremony honoring the ruler for his virtue and his astronomers for their expertise had to be rescheduled. An error of a full day also affected auspicious ceremonies tied to the first day of the month, season, and New Year.[5]

It was fortuitous for the Jesuits who arrived in China in the 1580s that their order had helped resolve one of the major calendrical problems bedeviling the Church in Europe, namely the controversy over the date for Easter. Pope Gregory XIII (1502–1585) had appointed the German Jesuit Christoph Clavius (1538–1612), a leading mathematician and churchman, to the Gregorian reform commission. With the promulgation of the Gregorian calendar in 1582, the Church had ascertained the proper date for Easter. Astronomers such as Clavius were already aware of Copernicus's (1473–1543) sun-centered (heliocentric) challenge to the earth-centered (geocentric) model of the solar system. The earth was a planet in Copernicus's system, and it orbited around the sun with the other planets, as well as rotating daily on its own axis.[6]

Clavius prepared a series of mathematical textbooks at the Roman College, beginning with his annotated edition of Euclid's *Ele-*

ments of Geometry. Matteo Ricci later used this edition for the classical Chinese translation of Euclid that he helped to prepare in 1607. Knowledge of geometry, especially the first four or six books of Euclid, became particularly important in sixteenth-century Europe for artisans, cartographers, architects, land surveyors, and painters, while in the late fifteenth century expertise in geometry had become increasingly valued for fortifications and ballistics.[7]

After three hundred years of adding intercalary months, the discrepancy between the Chinese lunisolar calendar of twelve lunar months of twenty-nine to thirty days each and the solar year approximation of 365.2425 days became significant. The newly arrived Jesuits were well equipped to participate in late Ming calendar reform because of their training in Clavius's mathematical astronomy.[8]

Until officials associated with the Jesuits appealed to the emperor for a substantial reform of the Ming astronomy system in the 1630s, the government was content to rely on special supervising officials to recruit new talent for the bureau. The reform, however, opened the door for leaders of the Ming and Qing dynasties to accept Jesuits as calendrical experts, just as earlier rulers had accepted Indian, Persian, and Muslim specialists. To gain support for the missionary enterprise, Ricci decided to present the Jesuits in China as specialists in mathematical astronomy. This plan was worked out in advance with Xu Guangqi (1562–1633) and Li Zhizao (1565–1630), then high officials in the Ming government. Li, for example, studied European mathematics and astronomy, in addition to geography, after meeting Ricci in Beijing in 1601.[9]

Reformers presented a Sino-Jesuit calendrical system in stages between 1631 and 1635, but the final steps of implementation failed, in part due to Xu's death in 1633 and the fall of the Ming dynasty in 1644. By 1628, Li Zhizao had compiled the *First Collection of Celestial Studies,* which included Xu's and Ricci's *Elements of Geometry.* Li's collection represented the last stage of Ptolemaic astronomy in

China. It would be replaced in the 1630s by the earth- and sun-centered (geoheliocentric) system devised by Tycho Brahe (1546–1601) at the end of the sixteenth century. In the face of the Copernican challenge, this new model, called the "Tychonic system," represented a compromise position beyond Ptolemy's increasingly discredited geocentrism, but one not yet acknowledging Copernican heliocentrism.[10]

The Jesuits and their collaborators also translated many works of Aristotelian natural philosophy into classical Chinese. These translations presented the system of schools, curricula, examinations, and degrees in a Thomistic manner and were based on both Jesuit pedagogy at the Roman College and Portuguese texts, which represented a new synthesis of the works of Aristotle and the interpretations of Thomas Aquinas (1225–1274).[11]

For the future of calendrical reform, however, the most significant Jesuits who arrived in China in 1621 were the German Johann Adam Schall von Bell (1591–1666) and the Italian Giacomo Rho (1593–1638). Both continued the geoheliocentric work initiated by Brahe and compiled by Li, in which the sun and its planets revolved around the earth.[12]

Late Ming Mathematical Astronomy

In 1629, Xu Guangqi, Li Zhizao, and their coterie of Jesuit translators produced an essential series of translations on astronomical reform from the latest European works. They entitled it the *Mathematical Astronomy of the Chongzhen Reign* (1635), and although it was never published as a complete collection, many of its included works were published. Some of these had been translated before 1629, such as Li and Schall's book on the telescope, *On the Farseeing Optic Glasses*, completed in 1626.

To avoid the appearance that European learning took precedence over Chinese learning, Xu stressed to the emperor that Western methods would yield a comprehensive system that would harmonize European and Chinese measurements: "Melting the material and substance of Western knowledge, we will cast them into the mold of the Grand Concordance system." Such accommodative language diverted attention from what the translations actually signified, that is, a new cosmological system. Unlike Li Zhizao, who earlier in 1610 had stressed the newness of Jesuit calendrical methods, Xu in 1631 minimized novelty and stressed accuracy. This emphasis on unifying European and Chinese knowledge was a tactic that would continue until 1900.[13]

In 1644, soon after the Ming dynasty was overthrown, Schall introduced the works included in the *Mathematical Astronomy of the Chongzhen Reign* to the conquering Qing court. In addition, when the Manchu Shunzhi emperor (r. 1644–1661) appointed Schall as the first European head of the Qing Astrocalendrical Bureau, Schall strategically changed the title of the collection to *Mathematical Astronomy According to New Western Methods* and had it recopied in 1645 in 103 volumes with minor rearrangements. The Qing dynasty's new Temporal Model astronomy system accordingly was based on the European system adapted for the vanquished Ming dynasty.

The Jesuits' new loyalty to the conquering Qing did not please literati loyal to the fallen Ming, who had been shut out of the new government. Indeed, this episode confirmed the literati's view of Schall and the Jesuits as opportunists, particularly since Schall had also been eager to work with the anti-Ming rebel leader Li Zicheng (1605?–1645) when Li had briefly captured Beijing in 1644 and precipitated the crisis that led to the Ming fall. Indeed, some literati were to argue to the Shunzhi court in 1660 that Schall had subordinated the Chinese empire to Europe, and in particular accused Schall and the Jesuits of touting the superiority of Christianity through their calendar. After Schall became director of the bureau, except for a short period from 1664 to 1669 when the Jesuits were successfully bumped from their positions, a Jesuit was in charge of the bureau from 1645 until 1775, although as we will see, the society was suppressed by the pope in 1773.[14]

From the "Vaulted Heavens" to the Tychonic World System

Since the early Han period (206 B.C.E.–220 C.E.), two ancient Chinese models had shaped Chinese thinking about their place in the cosmos. According to one, the "vaulted heavens" *(gaitian)* cosmology, the heavens arched over a flat, square earth like a hemispherical dome, or an umbrella-like canopy. Its classical alternative, beginning in the transition from the early to later Han, was called the "spherical heavens" *(huntian)* cosmology. In this view of the universe, an egg-white-like cosmos surrounded the yellow yolk-like earth. This view became influential between 100 and 180 C.E. but was not further elaborated.[15]

During the Ming and Qing dynasties, the Copernican system replaced the Tychonic system in Protestant European astronomy, but in China the Tychonic system continued to be used, hampering advances in astronomy. The Jesuits failed to introduce the Copernican system in a timely fashion, even though, for example, a few of Galileo's discoveries (albeit not his support of heliocentricity) were noted in Ming Jesuit translations. Nevertheless, the Tychonic age in the Astrocalendrical Bureau meant that by the 1630s Chinese specialists had available to them a rich toolkit of new computational techniques, more accurate observations, a new view of the cosmos, and the latest precision instruments.[16]

Numerical Tables, Star Maps, and Instrumentation

Xu Guangqi recognized how important European astronomical tables, maps of the heavens, and instruments were for calendar reform. From 1631 to 1635, the Chinese created correspondence tables between European calculation tables and Chinese measurements.[17] In addition, the Jesuits introduced trigonometric tables and translated in 1624 *Arithmetica Logarithmica* by the Englishman Henry Briggs (1556–1630). Jean Nicholas Smogolenski (1611–1656) and his collaborator Xue Fengzuo (1600–1680), too, introduced trigonometric logarithms in the early Qing. The use of both arithmetical and trigonometric logarithms enhanced astronomical calculations.[18]

In addition, the Jesuits provided the Ming court with up-to-date star maps, which culminated a classical tradition in China dating back to the *Classic of Stars* around 70 B.C.E. The *Catalog and Atlas of Fixed Stars* in the late Ming improved on the mapping of fixed stars completed earlier in China. Yet because the Jesuits were unable to correlate all the stars known to them with the Chinese star maps (and thus could not supplant the Chinese system of constellations with Western constellation names), the Chinese system was kept.

The reformers were enthusiastic about using the quadrant for measuring the altitude of celestial bodies, the parallactic ruler for measuring time, the celestial globe, the sextant, the equatorial astrolabe to observe the movement of the stars, the ecliptic or an equatorial armillary sphere to measure planetary motions, and gnomons to find the declination of the sun through the year (Figure 1.1). The telescope gave a special boost because it allowed study of the phases of Venus and the motions of the four Jovian moons around Jupiter—it was used as well to observe a solar eclipse in China on October 25, 1631. The German astronomer Johannes Kepler (1571–1630) also made a wooden pinhole device to measure the apparent diameters of the sun and moon and to observe solar eclipses in 1600, a device that was introduced into China circa 1632.[19]

The Jesuits also became the Qing court's clockmakers, and a Jesuit, usually Swiss, was placed in charge of the emperor's clock collection. Later, under the Kangxi emperor, Chinese clockmakers who worked under the Jesuits made native mechanical clocks. Im-

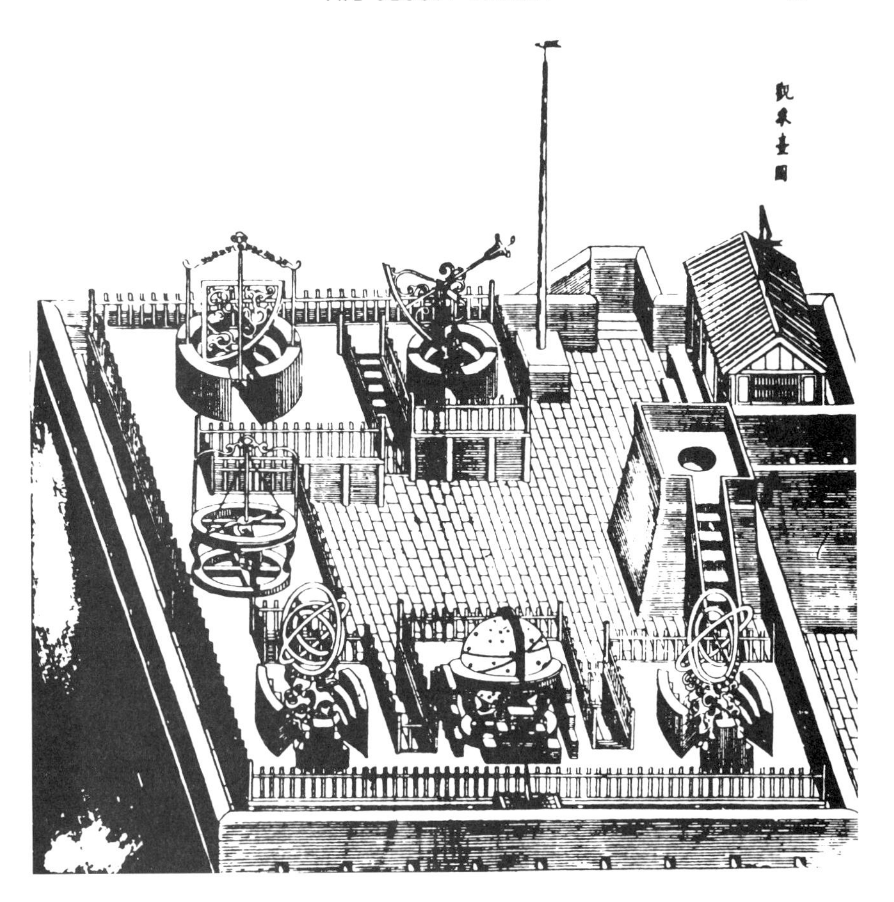

Figure 1.1. Verbiest's instruments in the Beijing Observatory. *Source:* F. Verbiest, *Xinzhi yixiang tu* (Beijing, 1674).

perial clock-making workshops were also established in Suzhou and Hangzhou. By the eighteenth century, both telescopes and mechanical clocks were common playthings in the imperial palace.[20]

The Jesuits in China were cut off from the latest developments in European science and technology, and they did not keep pace with advances in astronomical instrumentation and precision timepieces. But perhaps more significantly, they also neglected to realize the revolutionary implications of analytical geometry and the differential and integral calculus in the eighteenth century. Because Jesuit mathematicians stressed a static geometric mathematics, they never expanded the curriculum in the Jesuit colleges beyond Euclidian geometry. For the Jesuits, only Aristotelian natural philosophy, not mathematics, could explain movement.[21] After the Church proscribed Copernicanism in 1616 and condemned Galileo in 1633, the Catholic Church and its leading intellectuals in Europe also rarely forged ahead in *scientia* and its new instrumentation.[22]

If the Jesuits were not communicating cutting-edge advances in science, what did they do for the Chinese government? The Jesuits provided the imperial government with a new cosmological system, mathematical tables, and instrumental techniques needed to satisfy the carefully itemized list of demands that Xu Guangqi had outlined in 1629. This approach, as letters from Jesuits back to Europe indicated, sanctioned the religious success of the Jesuits in China.[23]

Jesuit Translations of Scientia in Classical Chinese

Between 1592 and 1606, the Portuguese Jesuit College of Coimbra produced a series of Latin textbooks and commentaries for the required works of Aristotle, which the Jesuits brought to China. In their translations, the Jesuits and their converts frequently stressed the syllogism as the tool to pursue *scientia*, which was a mark of specialized learning. To communicate and overcome the difficulties of translation, the Jesuits talked about their knowledge systems in classical Chinese terms that Ming literati chose. The translated terms and transcriptions that were adopted suggested correspondences and parallels when Chinese terms were commensurable.[24]

Transcriptions were necessary in some cases. To convey the meaning of *philosophia* in Western learning to literati, for example, Giulio Aleni (1582–1649) first presented a transcription using Chinese characters for their sounds: *feilusuofei*. With the help of his

Chinese adepts, he then linked this transcription to Chinese terminology for philosophy. Aleni had also studied under Clavius. His *Summary of Western Learning* (1622) presented a classification of the sciences that corresponded to the general Western European standards of the sixteenth century, which attributed a more important place to mathematics and technology.

Alfonso Vagnoni (1566–1640) distinguished between the ancient *scientia* of the Chinese and the new *scientia* of Europe in light of the investigation of things and extension of knowledge. Similarly, Francisco Sambiasi (1582–1649) translated philosophy as the "study of investigating things and extending knowledge" and the investigator as "one who investigated things." For physics, the Swiss Jesuit Johann Schreck (also known as Terrenz, 1576–1630) prepared his *Diagrams and Explanations of the Marvelous Devices of the Far West* (1627) with the help of Wang Zheng (1571–1644). It complemented the *Mathematical Astronomy of the Chongzhen Reign* by introducing Renaissance mechanics, machine technology such as clocks, and topics such as optics and heat.[25]

Acceptance of a portion of European *scientia* was possible in Ming and Qing China precisely because Chinese literati themselves had longstanding and commensurable interests in natural phenomena, which had undergone a significant revival in the sixteenth century before Europeans arrived in South China.

Sino-Jesuit Accommodations

One aspect of the accommodation strategy that characterized the Jesuit-Chinese interaction in the seventeenth century was the joint use of Chinese terms such as the "investigation of things" *(gewu)* and "exhaustively mastering principles" *(qiongli)*. Such expressions indicate some of the commensurable concepts that were at the core of the intellectual encounter between Ming-Qing China and early modern Europe. Mutually understood terms dealing with *scientia* and natural studies suggested to both Chinese literati and the Jesuits intellectual correspondences between Catholic Europe and Ming China.

For example, early Jesuit translations into classical Chinese of Aristotle's theory of the four elements in Vagnoni's *Treatise on the Composition of the Universe* (1633) and Agricola's *De Re Metallica* (1640) both used in their titles the themes of investigating things and extending knowledge in light of the Latin *scientia.* And each side, Chinese and Jesuit, sought to wipe away differences by claiming complete concordance, when in fact each aimed to achieve diametrically opposed results. The Jesuits effaced the Chinese content of the investigation of things with Western European natural studies as a way of enabling the Chinese to know heaven and accept the Church. The Chinese, in turn, effaced Western learning with native traditions of investigating things and extending knowledge, which allowed them to assert that European learning originated in China.[26]

Ricci also had an agenda: he thought that mathematics—that is, a Euclidian mathematics that was reduced to a set of fundamental, if preliminary, procedures and principles—would prepare the Chinese for the higher truths of Christianity. His use of logic was intended to convince Chinese of the necessary existence of God.[27]

Five Phases versus Four Elements

Both sides saw an order and purpose in the cosmos. The Jesuits' worldview encompassed theology and geography to unify God and nature. Most Chinese literati also saw the earth and heavens as a harmonious whole, but they framed arguments for the design of the cosmos around an eternally evolving Way (*Dao*) rather than around the linear chronology of a divine providence. In place of a cosmos made up of four elements (air, fire, earth, and water), the Chinese conceived of change in light of an undifferentiated Supreme Ultimate with no opposites. It spontaneously split into a completely differentiated stage of yin and yang. The interaction of yang and yin on the stuff (qi) of the universe then set in motion the five phases (earth, fire, metal, water, and wood) of cosmic change and yielded the concomitant production and destruction cycles of the myriad things in the world.[28]

Vagnoni's *Treatise on the Composition of the Universe* was in part a refracted presentation of the theory of the four elements from Aristotle's works. In his translation, for example, Vagnoni vainly tried to convince the Chinese of the error of their ways for including wood and metal and excluding air as the building blocks of things in the world. His use of translation for the Jesuit encounter with Chinese literati built on Ricci's earlier efforts to expound a theory of the four elements in 1595.[29] Vagnoni coined the new term for a primary element to render Aristotle's concept of element into classical Chinese and to demarcate it from the Chinese notion of five phases:

> An element is a pure substance in that when divided it does not form any other sort of thing. It can only form a composite thing made of several sorts. What is a pure substance? It is a substance of a single nature with no other composite elements. Accordingly, the myriad things in the world are distinguished by their being pure or composite. The pure elements are the four elements of earth, water, air, and fire. Composite things take on five forms such as the category of rain, dew, thunder, and lightning, the category of metals and stones, the category of plants, trees, and the five grains, the category of birds and animals, and the category of humans. All five of these forms are composites of the four elements.[30]

Vagnoni's concern from the outset was to gainsay Chinese inclusion of metal and wood as elements. Replying to a query by a Ming literatus, Vagnoni denied the very existence of five phases:

> If one observes how the myriad things are formed, it is not from metal and wood. The many categories of humans, reptiles, birds, and animals are examples of this. Therefore, metal and wood cannot be the primary elements of the myriad things. Moreover, who does not know that metal and wood are in actuality themselves composites of water, fire, and earth. As composites they cannot be primary elements. There are many additional composites that might be called primary elements, such as plants, stones, and other things, and which could be arrayed as primary elements. Consequently, they are not limited to just five. Why only select metal and wood?[31]

In the dialogue that Vagnoni imagined with a Chinese, there was an assumed agreement that at least three of the Chinese five elements—namely, water, fire, and earth—overlapped with the four elements of Aristotle. But Vagnoni believed that the Chinese had miscategorized the elements by failing to ascertain that in addition to water, fire, and earth, air was the only other pure substance. Vagnoni and his Chinese informant made no effort to conceptualize the five phases in Chinese terms, which described the five phases not as substances but as marked phases in every sequence of change, whose configurations over time were the key to the delineation of the myriad things in the world.

Of particular importance to the Jesuits in their efforts to promote Aristotle's views of nature and principles was their belief that air was one of the four primary elements. This entailed an equation of the Chinese notion of qi as the stuff of the world with the Aristotelian element of air. Interestingly, Aristotle had complemented the four elements with ether as a fifth substance to account for celestial bodies that were not susceptible to change and remained in eternal, circular motion. The sublunar world was composed of the four elements and subject to finite motions of all kinds in a world of constant change and transformation.

The Chinese notion of qi also worried Ricci and his colleagues because they saw in it a materialist continuum that encompassed all matter. In other words, qi verged on materialistic pantheism, leaving no place for the unlimited spiritual power of God. Consequently, Vagnoni tried in his dialogue to equate qi and air, thus demoting it to one of the four pure physical substances. The conversion of qi from an unlimited presence in all physical and spiritual things into a single substance making up some but not all things was based in part on what Vagnoni perceived as the strictly physical notion of qi. Rather than considering them building blocks of the universe, Chinese literati perceived in the five phases evidence for the successive changes in all things. Rather than the lifeless element air (or ether) enunciated by Vagnoni, Chinese perceived in qi a more fundamental material, vital, and spiritual unity, which pervaded all things in the cosmos and undergirded the space-time changes in yin and yang.[32]

Some late Ming literati thought that qi, when it took physical shape through congealing, made up everything between heaven and earth. For them, qi could be explained as both organized activity and the material structures that embodied it. According to this view, the five phases actually referred to qualities of qi that made up all substances—another side of their denotation as five sorts of fundamental processes. For Song Yingxing (b. ca. 1600), who wrote on traditional technology, two of the five phases were basic ingredients of matter but were not elements.[33]

The Jesuits and Mappa Mundi in Ming China

Jesuits added to the geographical knowledge of the Ming literati in the late sixteenth century. For instance, the first translated edition of Matteo Ricci's map of the world *(mappa mundi)*, which was produced with the help of Chinese converts, was printed in 1584. A flattened sphere projection with parallel latitudes and curving longitudes, Ricci's world map went through eight editions between 1584 and 1608. The third edition was entitled the *Complete Map of the Myriad Countries on the Earth* and printed in 1602 with the help of Li Zhizao (Map 1.1).

The map showed the Chinese for the first time the exact location of Europe. In addition, Ricci's maps contained technical lessons for Chinese geographers: (1) how cartographers could localize places by means of circles of latitude and longitude; (2) many geographical terms and names, including Chinese terms for Europe, Asia, America, and Africa (which were Ricci's invention); (3) the most recent discoveries by European explorers; (4) the existence of five terrestrial continents surrounded by large oceans; (5) the sphericity of the earth; and (6) five geographical zones and their location from north to south on the earth, that is, the Arctic and Antarctic circles, and the temperate, tropical, and subtropical zones.[34]

One indication of the influence of Ricci's maps, especially the 1608 edition printed in Beijing, was that many late Ming literati scholars included it in their geographical works. In addition to Ricci's 1584 map of the world, Ming scholars often also included European depictions of the northern and southern hemispheres, along with traditional maps of the four seas.[35]

The only known copy of the first Chinese map of the world, created in 1593, was entitled the *Comprehensive Map of Heaven and Earth and the Myriad Countries and Ancient and Modern Persons and Artifacts.* It included geographical information from Ricci's map of 1584, which is now lost. The 1593 Chinese map offered a traditional representation of China with foreign lands arranged around the periphery. It served chiefly as an administrative map for officials and thus included statistical information such as population (conventionally based on families) and locally produced commodities. Topological rather than topographical, the 1593 Chinese version fit European lands in along its edges without affecting tra-

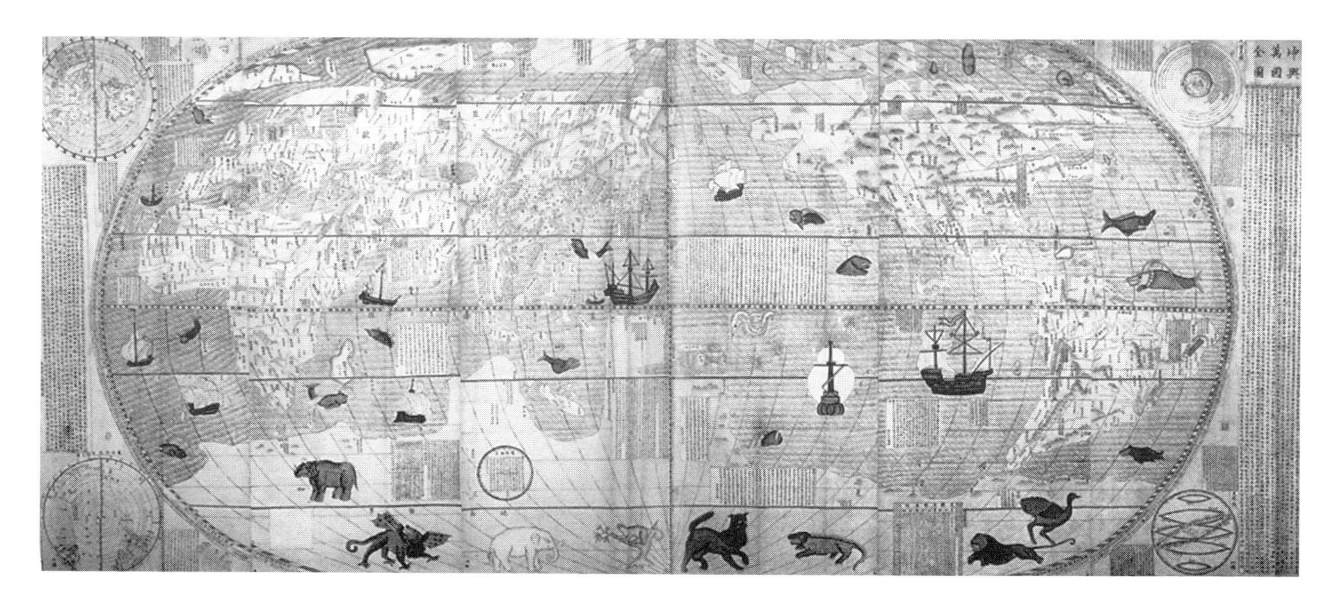

Map 1.1. Complete Map of the Myriad Countries on the Earth (Kunyu wanguo quantu).
Source: http://geog.hkbu.edu.hk/GEOG1150/Chinese/Catalog_main_11.htm.

ditional cartography. The New World was shown as a series of small islands surrounding China.[36]

Aleni and Yang Tingyun (1562–1627) quickly followed with an account of foreign countries, which was later included in the 1628 *First Collection of Celestial Studies* compiled by Li Zhizao. Aleni's translation represented the first detailed exposition in Chinese of a world geography that drew on Renaissance traditions of lore and cosmography to describe Asia, Europe, Africa, and the Americas. Another section focused on the oceans. In addition to the *First Collection of Celestial Studies,* other Chinese collectanea reproduced Aleni's treatise in the eighteenth century, thus making it more influential during the Qing dynasty than Ricci's maps, which were quickly forgotten by literati.[37]

Although Ricci introduced the system of longitude and latitude to Ming China, schematic grid maps, such as the *Map of the Tracks of Emperor Yu* carved in 1137 (Map 1.2), still exercised a dominant influence on Chinese cartography throughout the late imperial period. The Chinese grid style treated maps and texts as integral rather than independent elements. The Ming loyalist Huang Zongxi (1610–1695) wholeheartedly embraced this style when he initiated in 1673 an admired genre known as Complete Maps of All under Heaven, which was continued without making the grid explicit by a number of talented literati scholars interested in mapmaking.[38]

In the early Qing, Ferdinand Verbiest (1623–1688), with the help of others, produced two works that dealt with world geography. Based largely on Aleni's work, Verbiest's 1674 world atlas included a comprehensive map in two hemispheres with gazetteer information about each part of the globe. Similarly, Verbiest's *Essential Records about the West* drew on the topical organization in Aleni's 1637 *Answers about the West,* which compared China to the Europe in light of geographical lore. The compilers of the Imperial Library catalog during the 1780s considered these works important enough to copy them into the collection.[39]

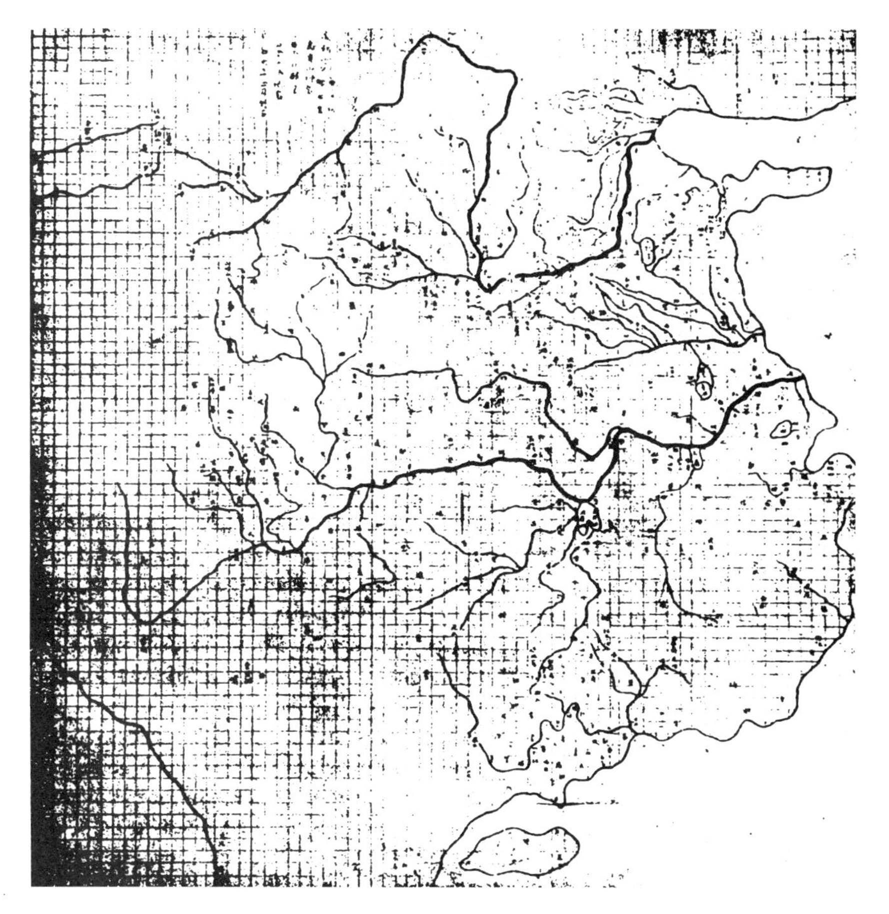

Map 1.2. Map of the Tracks of Emperor Yu (Yuji tu). Source: Reprinted from Joseph Needham, *Science and Civilisation in China,* vol. 3, Cambridge University Press, 1959. Copyright © 1959. Reprinted with the permission of Cambridge University Press.

The final stage in the development of traditional Chinese map-making came between 1708 and 1718 when French Jesuits, on behalf of the Kangxi emperor, systematically surveyed the entire Manchu realm. They drew up a series of maps of the Qing empire and its border areas, which became the 1718 *Kangxi Atlas.* Along with succeeding maps in the Yongzheng (1723–1735) and Qianlong (1736–1795) periods, the *Kangxi Atlas* surpassed earlier Jesuit surveys. The Manchu court restricted access to and local reproduction of these surveys and maps. Nevertheless, after it was brought to Europe the *Kangxi Atlas* remained the chief source of information about China's geography until the twentieth century.[40]

The Breakdown of Christian Consensus about China

Conflicts among Catholic missions worldwide were exacerbated by a longstanding competition between Spain, which sponsored the Dominicans, Franciscans, and Augustinians, and Portugal, which trained and sent the Jesuits abroad. Later, too, French Jesuits were accused of glorifying French science. Each denomination followed different strategies to convert the peoples of the world and please their patrons. Initially, the better educated and more urbane Jesuits put in place an accommodation policy to facilitate their dealings with knowledgeable Chinese literati and the sophisticated courts of the Ming and Qing empires. This policy was not accommodation at any cost, however. Ricci refused to baptize converts who clung to their concubines or approved homosexuality. He also attacked orthodox metaphysics as atheism.[41]

Pope Clement XI (1649–1721), unlike his predecessor, sympathized with the purists in the Catholic Church, although before 1702 he did not take sides. In 1701, however, he appointed an opponent of the Jesuits as his legate to China to ensure that the prohibition of Chinese rituals among Chinese converts was enforced. Disenchanted with the pope's representatives, the Kangxi emperor decided to deport all missionary troublemakers. To the dismay of the Jesuits, the new policy also rolled back the emperor's 1692 Edict of Toleration. After 1706, he required all missionaries to have an imperial certificate to maintain residence in Qing China.[42]

The Kangxi emperor now understood that the Jesuits were

agents of Rome. The eldest prince, for his part, thought the missionaries were spies for the Spanish or Portuguese who were preparing an invasion of China. In 1717, perhaps as a result of such fears, the emperor prohibited all missionary work in China. On the papal side, the Decree of the Inquisition signed by Clement in 1710 imposed an absolute ban on writing on the Rites issue. Those who failed to heed the injunction, particularly the Jesuits, would face excommunication. Later papal bulls of 1715 and 1742 silenced all discussion of the controversy in Europe and cemented the triumph of the anti-Jesuit clergy.

Starting after 1705, the influence of the Jesuits began to decline in China and across Europe. In the late 1720s, the Kangxi emperor understood that the pope could not command the Episcopalian English or Protestant Dutch and that the Jesuit claim that their order represented the strongest ruler in Europe was false.[43] The Rites and Terms Controversy also backfired on the Jesuits in Europe. And although in 1749 there were 22,600 members of the Society of Jesuits and some 669 Jesuit colleges and 176 seminaries worldwide, after 1773, the pope sequestered the society's property, and in 1764 its rich library collection in the Jesuit College of Clermont was sold. According to the 1773 decree, in which Pope Clement XIV dissolved the Society of Jesuits, the key issue was the Jesuit accommodation policy; it was unacceptable for the Jesuits to have placed China and its rituals on equal footing with Europe and Christianity.[44]

RECOVERING THE CHINESE CLASSICS

In the aftermath of the Rites Controversy, in which Chinese ancestor worship was debated in Europe, the Kangxi emperor sought to counter what he now considered undue Jesuit influence on the state's mathematical astronomy. In 1712–1713, he established the Studio for the Cultivation of Youth in the suburban Lofty Pavilion Garden-Palace to encourage astronomical work by the emperor's own Chinese, Manchu, and Mongol mathematicians (even though he still needed to employ French Jesuits to improve the calendar). As a result of the emperor's efforts, the social standing of mathematicians under the Qing rose in the eighteenth century, and literati learning increasingly valorized natural studies, mathematics, and medicine as objects of study.

Eighteenth-century Chinese scholars also restored ancient medical and mathematical classics, and through their antiquarian research and painstaking identification of forgeries, stressed the importance of rigorous analysis and impartial evidence drawn from ancient artifacts and documents. In short, they made verification a central concern for an emerging empirical theory of knowledge.[1] The critical community of evidential research scholars that was formed during the period is reminiscent of earlier European scholars who in the twelfth century translated ancient Greek and medieval Arabic mathematical and medical texts into Latin and helped forge the scholastic synthesis that the Jesuits introduced in China as their natural philosophy. Yet the European and Chinese scientific cultures still diverged during the eighteenth century. While mid-Qing literati concentrated on recovering their classical mathematics and medicine, focusing on the distant past to overcome recent failures, the Europeans went beyond their ancient masters to make significant breakthroughs during this era.[2]

Traditional Chinese medicine also looked to the past for new ideas. Intellectual and therapeutic developments increasingly focused on "heat factor" illnesses as a new category of disease, distinct from perennial "cold factor" illnesses. In addition, a reemphasis on ancient texts such as the *Treatise on Cold Damage Disorders* in the eighteenth century stimulated the reexamination of pre-Song therapies for cold- and heat-factor illnesses. Instead of using warming medicines to counter perennial cold-factor disorders, physicians increasingly prescribed cooling drugs and methods for infectious illnesses that brought on intense fever.[3]

Kangxi and the Academy of Mathematics in Beijing

The Kangxi emperor formed his Academy of Mathematics based loosely on the Parisian Academy of Sciences, but it was named after the Tang dynastic school for mathematics. The academy was established in 1713 in the Studio for the Cultivation of Youth, but only Qing literati and Manchu bannermen were appointed to it. No Jesuits were allowed in this inner coterie of imperial scholars, which included the third prince, Yinzhi (1677–1732). This post–Rites Controversy policy ensured that the Jesuits would not excessively influence court mathematics. The emperor now wanted his own men in charge of such technical knowledge.[4]

Li Guangdi (1642–1718), a leading patron of native specialists on mathematical astronomy in the Kangxi court, also avoided relying on the Jesuits. Li's group included Wang Lansheng, who had been granted the highest civil service degree in 1721 by the emperor because of his mathematical abilities and called a "palace graduate in mathematical astronomy." Wang entered the Studio for the Cultivation of Youth and helped the French Jesuits compile newly translated works on mathematical astronomy.[5]

The emperor recruited more than one hundred scholars—regardless of their civil examination status—to join the Academy of Mathematics. In addition to those in the academy who studied mathematics, astronomy, and music, a large number of instrument makers were hired for the technical needs of the new academy. A team of fifteen calculators verified the computations by using theoretical notions, mathematical techniques and applications, and numerical tables. Patterned after mathematical textbooks used in Jesuit colleges, the new works that were translated also introduced European algebra.[6]

After 1723, no further European mathematics was introduced into China until after the First Opium War (1839–1842). Notably missing in China was the European discovery of the calculus. Moreover, the incomplete version of Euclid's *Elements* remained the official version until 1865. French Jesuits, for example, tried to introduce Descartes's new algebraic notational forms in a treatise prepared for the emperor, but the emperor could not grasp the benefit of such general symbols that described the algebraic process. He preferred the calculating craft of algebra that quickly lead to the solution. Thus by 1730, the astronomical books of the Kangxi era were out-of-date by European standards.[7]

Kangxi-Era Computational Astronomy

Kangxi's efforts at reform culminated in the promulgation in 1724 of the *Compendium of Observational and Computational Astronomy.* The European astronomy in the *Compendium* was mostly a century old, but its 1742 *Supplement* adapted more recent European discoveries for traditional calendar reform. Such new techniques were wedded to calendar reform and not developed further, however. The *Compendium* did lead to Chinese observations and computations that no longer depended on Jesuit help, but mastery of Western methods by Chinese and Manchu specialists mostly demonstrated the observational and theoretical shortcomings of Chinese methods.[8]

Indeed, Qing specialists had no domestic incentive to go beyond the immediate needs of the Qing calendar, now successfully reformed. Nor did the Jesuits press them to do so: by 1725, the Jesuits were themselves no longer on the cutting edge of the early modern sciences, and their mathematics, no more than simple algebra, trigonometry, and logarithms, had been successfully adopted by late Ming and early Qing specialists. In the eighteenth century, a larger community of Qing classical scholars associated with textual studies would restore traditional Chinese mathematics to a level of intellectual prestige commensurate with the Jesuit version of European studies. Meanwhile, the Kangxi emperor's private sessions with the French Jesuits included training in surveying, mensuration, calculating, astronomy, geometry, and logical argument.[9]

When compared to eighteenth-century developments in Europe, the fate of the Qing Academy of Mathematics is instructive. In France, the Paris Academy of Sciences helped create a growing community of men of science and advanced the institutions that supported them. Such institutional changes encouraged the eclipse of general learned societies by more technical institutions. The establishment of strictly scientific disciplines in Europe by the late eighteenth century was accompanied by an expansion of universities and research institutes where professionalized science slowly incubated in institutions of higher learning, and specialized laboratories eventually replaced gentlemanly academies. Such developments in China only occurred in the late nineteenth century.[10]

The Kangxi Emperor and Mei Wending

As Jesuit influence in the court faltered, Manchu emperors increasingly patronized their own imperial mathematicians with questions about astronomy as a way of lessening the dynasty's dependence on foreign experts. In particular, the social status of the native mathematicians rose dramatically after 1700 when Han Chinese literati officials close to the Kangxi emperor increasingly patronized them. Mei Wending (1633–1721) was one such native mathematician who challenged the Jesuit monopoly in the Astrocalendrical Bureau and at court. In turn, the emperor enhanced Mei's astronomical credibility empirewide by according him unprecedented honor as a model for court mathematicians.[11]

After dealing with the astronomy crisis in the 1660s, the Kangxi emperor recognized the need for astronomical and mathematical expertise in court and in the bureaucracy. Although the Ming had banned public knowledge of computational astronomy and astrology in order to monopolize it in the Astrocalendrical Bureau, Kangxi opened them for official study by appropriate Han officials and trustworthy Manchu and Mongol army bannermen. His goal in part was to weaken the hold of certain Han and Muslim families that had dominated the bureau's transmission of mathematical astronomy. He also hoped to unseat the monopoly of entrenched Catholic converts in the bureau.[12] Because the bannermen regarded civil examinations and mathematical training as secondary to their military skills, however, their mathematical achievements were limited.

Overall, the Kangxi emperor's experts followed Mei Wending's lead in rejecting Jesuit efforts to insinuate Christianity into their astronomy. Mei's mathematical work fit well with the court's efforts to have a calendar that would fuse European and Chinese techniques into a comprehensive system and end wrangling on both sides. When the *Compendium of Observational and Computational*

Astronomy was drafted in 1722 and promulgated in 1724, for instance, it followed European models but was prepared by Chinese in the court with only indirect Jesuit input. The key figure was Mei Wending's grandson, Mei Juecheng (d. 1763), who had served on the staff since 1713 and had links to his grandfather's patron, Li Guangdi.

Mei Wending's reputation was so honored by the emperor that later accounts have tended to overlook the many Qing students of Chinese and European mathematical astronomy with whom he interacted. The work of Wang Xichan (1628–1682) remained in manuscript form, for example, and never had the influence of Mei's publications. Perhaps surprisingly, Mei Wending's mathematical career might also have gone unheralded. His father and other grandfather had been book collectors interested in numerology. As Ming loyalists, they kept their distance from the Manchu regime. In addition, although Mei recognized the value of Jesuit learning, he was leery of Christian influence in the Astrocalendrical Bureau.

During the emperor's southern tour in 1702, Li Guangdi presented the Kangxi emperor with a copy of Mei Wending's *Queries on Mathematical Astronomy,* written in the 1690s. Li hoped that the emperor would be pleased with Mei at a time when the Rites Controversy was forcing the court to rely less on the Jesuits for astronomical and mathematical expertise. After reading the work, the emperor was impressed, although he noted, perhaps after consulting the French Jesuits in court, that it had few mathematical problems.

Later in 1705, when the emperor again traveled south, he asked to meet with Mei Wending. For three days, Kangxi and Mei apparently discussed mathematical astronomy on the imperial barge, and the emperor was pleased that he had a native scholar who was as knowledgeable as the Jesuits. Mei also presented the emperor with a copy of his new work on trigonometry.

When the Jesuits proved to be unreliable intermediaries with the pope, the Kangxi emperor's discovery of Mei Wending and other Chinese mathematicians proved timely. Moreover, Mei's efforts to reconcile Chinese and European mathematical astronomy appealed to the emperor because it legitimated both traditions. After meeting with the emperor, Mei promoted the "Chinese origins" of Western computational astronomy. For instance, Mei traced the "vaulted heavens" cosmology of a square earth surrounded by rounded skies back to its classical alternative, the "spherical heavens" cosmology. In Mei's view, the celestial phenomena in the southern hemisphere could be derived from the same computational procedures used to demarcate the northern hemisphere.

Thereafter, the prominence of the Mei family in Qing mathematics was widely recognized. The Kangxi emperor secured the Mei family's eminence by declaring Mei Juecheng an Imperial School student in 1712, granting him a provincial degree in 1713, and promoting him to the highest degree as "a palace graduate in mathematical astronomy" in 1715. Mei was even allowed to take the palace examination without taking the (usually) required preliminary examination.[13]

Mei Wending and his family favored a quantitative approach to dealing with astronomical principles. Computations for grasping heavenly phenomena permitted numbers and principles to be linked:

> Someone might ask Master Mei: "Is mathematical astronomy a matter of concern for the classical scholar?" I reply: "Of course. I have heard that it is the classical scholar who masters the comprehensiveness of heaven, earth, and humans. Is it permissible to be enveloped by but not know the height of the heavens?"
>
> He might ask: "The scholar in knowing heaven knows only its principles. What use does he have of astronomy?" I reply: "Mathematical astronomy requires numbers. Outside of the

numbers there are no principles, and outside of the principles there are no numbers. Numbers are the orderly demarcation of principles. Numbers cannot be spoken of arbitrarily, but principles at times can be talked about via vague correspondences. Hence, arbitrary views have been associated [with principles], which have deluded the people and brought chaos to heaven's regularity. All this results from not obtaining the true principles and numbers and instead maliciously overturning reality."[14]

In Mei's view, the cumulative development of European methods was analogous to that in China; therefore the two traditions could be unified. For example, both European and Chinese scientists had demonstrated that the prediction of eclipses required knowledge of the time and position of celestial bodies in space, which they had developed beyond the limits of traditional calculation.

Another commonality Mei recognized had to do with geometry, which he regarded as the most systematic aspect of European studies and the root of its mathematics. Mei perceived geometry as an outgrowth of traditional mathematics—for example, he explained geometry by using traditional solutions for the sides of a right triangle. Indeed, his approach to computational astronomy was thoroughly geometrical. He seems to have thought that Euclid could be reduced to traditional mathematics. Even so, Mei Wending acknowledged that European spherical geometry provided new universals that the ancient Chinese sages had not anticipated.

When his grandson compiled Mei's collected works in 1761, forty years after his death, Mei Wending was recognized as the leading literatus who had mastered all the fields of European mathematics introduced to China. Moreover, until 1850, his *Complete Works on Mathematical Astronomy* was the starting point for efforts to reinvigorate traditional methods with the sophistication of new

European approaches. Robert Morrison (1782–1834) of the London Missionary Society, for example, had Mei Wending's *Complete Works on Mathematical Astronomy* in his library.[15]

Although he had rehabilitated ancient Chinese techniques, Mei was unable to locate all of the works included in the Ten Mathematical Classics because many were not widely available. And while mid-Qing scholars increasingly included mathematics in their classical fields of interest as a result of Mei Wending's influence, none of them followed up on later European mathematics. During the eighteenth century, no one transmitted to China new developments such as the kinematic solutions and moving models of the fluxional calculus of Newton (1643–1727) and the infinitesimal calculus of Leibniz (1646–1716), which directly linked time and motion. Consequently, Mei and his successors continued to rely exclusively on the static, essentialist geometry of Euclid and the qualitative motions of Aristotle.[16]

During the late eighteenth century, others took up the mathematical astronomy of Mei and the Jesuits. Evidential research scholars such as Jiang Yong (1681–1762), Dai Zhen (1724–1777), and Qian Daxin (1728–1804) wrote widely on computational astronomy, although on nativist grounds they ignored the more formulaic Renaissance science of algebra in favor of Chinese techniques.[17]

Mensuration and Cartography in the Eighteenth Century

The late Kangxi and Yongzheng bans on propagating European natural studies were highly ineffective at curtailing advances in geography and cartography. In the midst of Qing empire-building, as the court used Jesuit surveying techniques to measure its domains, evidential scholars quickly adopted and enhanced new discoveries from abroad. Civil officials and scholars in fact played an important part in the military expansion of the Qing realm during the late eighteenth century. Literati such as Zhao Yi (1727–1814) and Sun

Shiyi (1720–1796) gained valuable experience in the Burma campaigns (1766–1770), the annexation of Tibet (1790–1792), and the White Lotus uprising (1796), but they also faced hardships when assigned to Qing campaigns in Annam (1788–1789) and aboriginal revolts in Taiwan (1787–1788). While the Manchu dynasty took advantage of turmoil in Central Asia, Chinese literati took an inward turn intellectually by focusing on native topics in their new geographic works.

The Seventeenth-Century Turn Inward

Led by their naval commander Qi Jiguang (1528–1588) in the 1560s, hundreds of Ming coastal ships equipped with large European-style cannons destroyed the ships that Japanese pirates and their collaborators had used to maraud on land. Later, the Ming fleet joined forces with the Korean navy to resist Japanese invasions of the Korean peninsula in 1592 and 1597. Their combined forces of some five hundred ships and fifteen thousand men had superior tactics and technology and continually threatened Japan's land-based supply lines. Despite mobilizing a Japanese fleet of twelve thousand men on five hundred ships for the climatic battle at the Noryang Straits in December 1598, Toyotomi Hideyoshi (1536–1598) failed in his grandiose plans to use Korea as a stepping stone to conquer China. Some three hundred Japanese ships with ten thousand sailors were lost.[18]

Due in part to widespread travel throughout China, many scholars in the seventeenth century, particularly as a response to the Manchu military conquest in midcentury, returned to traditional questions of regional military strategy and local, coastal defense. Japan also entered a period of semi-seclusion. Meanwhile, the Dutch colonized Taiwan from 1623 when the Dutch East India Company contracted Chinese traders and farmers from Southeast China to settle the island.[19]

Subsequently Southern Ming loyalists led by Zheng Chenggong (Koxinga, 1624–1662) resisted the Manchus in major naval and land battles along the southeast coast in the 1640s and 1650s. Zheng's land and sea forces took heavy losses, however, when they moved up the Yangzi River to Nanjing in 1659, and were forced to retreat to Xiamen (Amoy). Southern Ming naval forces then challenged the Dutch garrison in northern Taiwan (called "Formosa" by Europeans) at Castle Zeelandia in April 1661 with a force of six hundred ships and twenty-five thousand sailors. The Dutch capitulated after a nine-month siege. Zheng demanded via a Dominican missionary that the Spanish in Manila recognize his suzerainty.[20]

For its part, the Qing government in 1662 ordered all Chinese coastal inhabitants to move inland to cut Zheng's supply lines and to negate the value of the coast as a battleground. Earlier, Manchu military forces had captured Ming cannons, which they had used to pound walled Chinese cities into submission while reorganizing their Chinese and Manchu bannermen into artillery units. The Manchus also turned to a naval fleet to defend the coastline, after Shi Lang (1621–1696), one of the Southern Ming's most capable admirals, joined the Manchus in 1646 because of a dispute with Zheng. Shi commanded Qing naval forces in the 1650s and 1660s along the Fujian coast. In July 1683, he led a Qing fleet of some three hundred warships and twenty thousand sailors, which subdued the Pescadores Islands. Taiwan fell to the Qing navy in October, and for the first time the island became part of imperial China.[21]

After Taiwan was annexed, Manchu emperors shifted their focus from sea to land, becoming preoccupied with the expansion of the Russians from Siberia into the Manchu homelands and the renewed dangers posed by the Zunghars in Central Asia. In addition, the Qing expanded its empire in Tibet and Turkestan. By the end of the eighteenth century, it had more than doubled the size of Ming China. When the First Opium War with England broke out in 1839, therefore, the Qing fleet was again mainly a coastal navy used principally for defense against pirates and local marauders.[22]

Because the Manchu conquest emanated from the northern steppe, Gu Yanwu (1613–1682), a leading voice of the early Qing turn toward precise studies, emphasized China's strategic positions vis-à-vis its traditional foreign neighbors to the north and northwest in his influential geographical treatise compiled between 1639–1662. Concerned with texts as much as with maps, Gu did not even mention Ricci's *mappa mundi*. He focused instead on the effects of topography on political and economic development within China. His findings, however, were based on his wide travels, careful firsthand observations, and the study of written materials. Likewise, Gu Zuyu's (1631–1692) study of historical geography, written from 1630 to 1660, explored native historical, administrative, and natural geography, emphasizing the importance of topography for military strategy. Literati disregarded Ricci's world map, and late Ming interest in maritime battles waned.[23]

Zuo Zongtang (1812–1885), however, was inspired by Gu Yanwu's and Gu Zuyu's geographical treatises to pursue what was to become a lifelong interest in Chinese topography and military strategy. His interest in Chinese Turkestan, in particular, underscored his insistence in the 1870s that troops be sent to Northwest China to prevent that area from falling permanently into Russian hands. Zuo's campaigns provoked the opposition of Li Hongzhang (1823–1901) and other Beijing officials who regarded naval power for coastal defense and protection of Korea from Japan as more pressing than the recovery of territory in the distant interior.

The compilers of both the official *Ming History* and the *Comprehensive Geography of the Great Qing Realm* had access to Ricci's and other Jesuit geographical works, but they dismissed as fictitious many of the Jesuits' claims and much of the information on their *mappa mundi*. Nevertheless, leading textual scholars served as editors of topographical material for the *Comprehensive Geography* project after 1687. The appointment of Gu Zuyu, perhaps the most qualified student of historical geography in his time, indicates the caliber of experts who carried out the *Comprehensive Geography* project.[24]

Similarly, the late Qianlong compilers of the 1787 edition of the *Comprehensive Analysis of Archival Sources during the August Qing Dynasty,* which included documents and materials covering the period 1644–1785, demoted the mention of Europe to a minor section on Italy within the category of the "Four Frontiers." The compilers dismissed the Italians as grandiose and argued that they were simply trying to impress Chinese with European customs, goods, governance, and education. Still, the *Comprehensive Analysis* included detailed geographic discussions of Europe and Russia.[25]

Moreover, the dramatic political influence that European surveying methods had had in China early in the eighteenth century piqued the interest of the Qianlong emperor and his court when the Kangxi map of the empire was updated using these methods. European geographical content may have been overlooked, but European methods were admired and copied in official geography as well as astronomy. Greater precision meant greater accuracy in mathematical astronomy, an expectation that the Jesuits used to enter and control the Astrocalendrical Bureau. Qing literati expected similar levels of accuracy in evidential learning, which drew from all three realms of classicism, geography, and mathematical astronomy.[26]

The reunion of classical and technical studies during the Qing also produced specialists who adopted an active, interventionist approach to problems ranging from river and flood control, to imperial statecraft, to the historicization of place names (and the discovery of forgeries through understanding how place names were derived historically), to land reclamation and hydraulic works, projects used to order physical space in the eighteenth century.[27]

More and more, diagrams and tables were used as aids in discussion, explanation, and classification. Geometrical diagrams were abundant in Mei Wending's writings, for example, because he used them to depict the mathematical nature of astronomy. In their attempts to comprehend celestial motions, Chinese astronomers shifted from strictly numerical procedures to geometric models of successive locations in space.[28]

Diagrams became for evidential scholars ingenious representations. The mathematization of the world, which in Europe was dependent on Newtonian mechanics and the calculus in the latter half of the eighteenth century, was unavailable to evidential scholars in China until the aftermath of the First Opium War—which meant that Qing scholars did not reconceptualize foreign lands.[29] The Chinese scholars did, however, combine evidential research methods with data collection and organization to make new advances.

For example, despite the turn away from concern with lands far from China, achievements in geographical knowledge during this period were evident in military defense and historical and descriptive maps, particularly those that described the Qing borders with Russia, Zungharia, and Kashgaria in Siberia and Central Asia. The mapping of Manchuria, too, under Jesuit direction, began circa 1690 in the aftermath of negotiations with the Russians to determine the boundaries of the Amur River in northeast Asia.

Such achievements lent themselves to the accumulation of geographical knowledge, which was made possible because evidential scholars stressed an empirical epistemology and focused on research topics that allowed for continuity in geographical research. As a result, geography emerged as a exacting discipline during the seventeenth century.[30] European cartographic technologies coexisted with earlier Chinese geographic practices such as cosmographic siting, landscape design, and urban arrangements. The geography of the Qing empire, too, was intertwined with the Manchu expansion into Central Asia, an expansion that required state-of-the-art mapping techniques from Europe to delineate the Russo-Chinese border.[31]

Cartography, Sino-Russian Relations, and Qing Imperial Interests

A crisis in Sino-Russian relations occurred from the 1670s to the 1690s, after the Manchus learned that the Russians had built a fortress in 1654 along the Amur River at Nerchinsk. The changing borders threatened the Manchu homeland, and the Kangxi emperor refused any further trade or diplomatic relations with Russia until the deadlock was resolved. Meanwhile, the Russians and Zunghar Mongols both expanded their interests in the northwest while the Qing were preoccupied in the south and southwest during the Revolt of Three Feudatories from 1673 to 1681. Much like late Ming calendar reform, Qing recognition of its geographic needs preceded European contributions to Chinese cartography.[32]

The lack of a clear boundary in the Amur River area and the ambiguous claims to sovereignty there led to the treaty of Nerchinsk in 1689, negotiated by the Jesuits, which demarcated the frontiers between the Qing and Russia. Later, during the Rites Controversy, the Manchu court was embroiled simultaneously in military threats from both Zunghars and Russians along the borders of the empire, challenges that introduced new elements into the storm over the Jesuits and their loyalty to the Qing dynasty.

Qing leaders commissioned the Jesuits and others to provide the map data required to stem Russian infiltration into Manchu and Mongolian homelands. In the process, the Kangxi court's awareness of the actual geographical divisions of the Sino-Russian frontier slowly caught up with their knowledge of Southeast Asia. In addition, the Kangxi and Qianlong emperors used new techniques for trigonometric surveying to map Qing dominions in the eighteenth century. Jesuits and bannermen surveyed the Manchu homelands between 1709 and 1712 and completed a map of greater Manchuria. The text and maps included in the 1733 edition of the *Collected Statutes and Precedents of the Great Qing* were concerned with military deployments and garrison towns. The maps that the Jesuits prepared for the Manchu homelands became the starting point for later Japanese and European maps of the region.[33]

The imperial ambitions of the Qing dynasty, Russia, and the Mongol Zunghars in Central Asia, as described by competing maps of the era, led to the redrawing of frontier boundaries between Russia and the Qing as well as the crushing of Zungharia in 1760 by Qing armies.[34] When Russia and China mapped their common borders in the treaties of Nerchinsk in 1689 and Kiakhta in 1727, both applied new surveying techniques.[35] The atlas and its subsequent Qianlong-era revisions had features like those of contemporary European maps: for example, they drew on astronomical observations to calculate precisely longitude and latitude. In the original atlas, however, China was presented as one distinct part of the Qing empire; the Manchu homelands were another. In addition, two other versions of the map, which had been created from the same surveys, were entirely in Chinese with no Manchu script to accommodate Han Chinese cultural sensibilities. Zungharia and Tibet were later added as distinct spheres of the Manchu empire.[36]

The Kangxi surveys were completed by 1717, and Qianlong finished revisions in 1755. (Similarly, the Russian imperial atlas appeared in 1745.) The *Kangxi Atlas* decisively changed how China was represented on European maps from the moment the Jesuit versions first arrived in France in 1725 and influenced scholars in Paris, London, and elsewhere.[37] Similarly, the latest mapping technology allowed the Qing to consolidate and legitimate the empire, and became the basis for all of China's territorial claims in the twentieth century.[38]

More importantly, by neutralizing Russia the Qing court prevented a Russo-Zunghar alliance against the Qing empire (Map 2.1). The

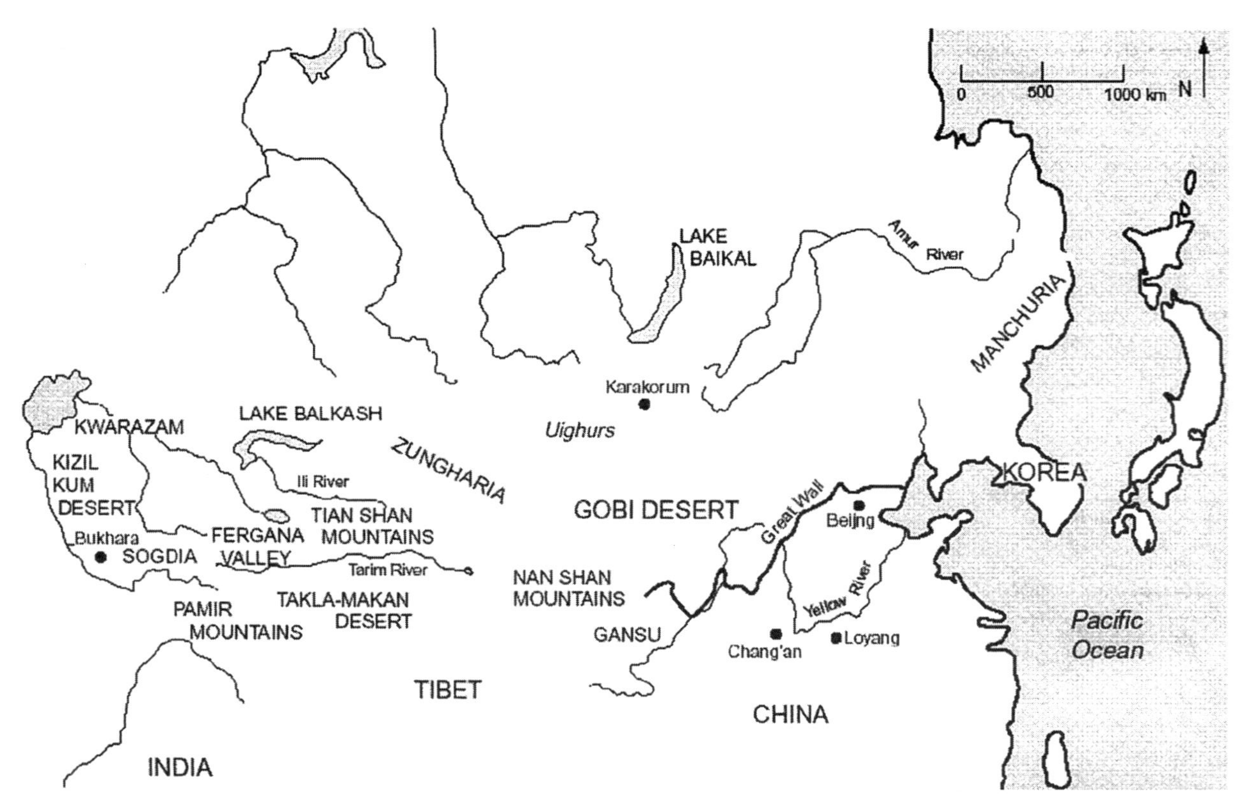

Map 2.1. Qing dynasty territories in proximity to Russian and Zunghar lands in Central Asia.

treaty of Nerchinsk represented a compromise in which the marking out of the frontier was favorable to the Manchus, while the Russians kept Nerchinsk. In addition, the Manchus conceded that trade could be initiated by either side, and each could cross the border with passports.

The economic concessions made by the Qing government in the 1689 treaty came just in time for the Manchus: the leader of Zungharia proposed an alliance with the Russians in 1690, but because the Russians were bound by treaty with the Qing, no partnership was made. Instead, the Kangxi emperor was left free to eliminate the threat posed by the Zunghars, arguably the last nomadic empire, which he did in 1696 by reducing the Mongols as a potentially divisive third force in Central Asia.[39] When the Zunghar threat revived in 1728, the Manchus again rallied, defeating the Zunghars in wars during the 1750s that were facilitated by the Treaty of Kiakhta.

The Kiakhta treaty ended Russian interference and established officially supervised trade in Amuria, which stabilized the Russian-Qing frontier until the nineteenth century. Diplomacy, warfare, and the timely mapping of strategic frontiers enabled the Qing dynasty to incorporate major portions of Central Asia at the expense of the Mongols, Uighurs, Kazaks, Tajiks, and Russians.[40]

Medical Works and the Recovery of Antiquity

During the Ming and Qing empires, the medical classics provided scholars and physicians with a set of general assumptions about the application of qi, yin and yang, the five phases, and the system of circulation tracts to understand the human body and its susceptibility to illness, which was defined as a loss of harmony in the body's operations. Circulation tracts of qi, for example, allowed doctors to map points on the skin to facilitate healing using acupuncture and moxibustion (the burning of mugwort, a small, spongy herb). Since antiquity, physicians had thought of the internal flow of qi through a series of main and branch conduits as being the body's vital current.[41]

For Qing literati-physicians, textual mastery of the medical classics and their commentaries was required to recover ancient principles and practice. The formation of evidential scholarship and the return to antiquity in medicine turned out to be mutually reinforcing trends.

Reconfiguring the Medical Classics

The oldest and most important Chinese medical classic was the *Inner Canon of the Yellow Emperor,* which was completed in the first century B.C.E. When set in its orthodox form during the Northern Song, it focused on anatomy, physiology, and hygiene in the part called the *Basic Questions,* while presenting a basic understanding of acupuncture and moxibustion in the *Divine Pivot.* Treatments using drugs were rare, and the focus was on preventative medicine. The two parts of the *Inner Canon* had been regarded since the Song as the basis for medical theory and practice.[42]

The later *Treatise on Cold Damage Disorders* by Zhang Ji (150–219) applied the *Inner Canon* to drug therapy. Zhang wrote his book in response to contemporary epidemics. Song literati considered the *Inner Canon* to be the fount of medical doctrine, for which the *Treatise* provided a guide to clinical practice. During the Northern Song, for example, the *Treatise* guided treatment of infectious diseases brought by the winds. These "cold damage disorders" were thought to be responsible for the increase in southern epidemics from 1045 to 1060.[43]

After Zhang Ji's original version, which had been called the *Treatise on Cold Damage and Miscellaneous Disorders,* was restored by the palace physician Wang Xi (Shuhe, 180–270), other revisions followed during the Northern Song, when it was divided into three separate books: the *Treatise on Cold Damage Disorders*, the *Canon of the Gold Casket and Jade Cases*, and *Essentials and Discussions of Formulas in the Gold Casket*. These works classified illnesses according to their chief symptoms and divided the symptoms into six stages, with three categories belonging to yang and three to yin. In the late Ming, scholar-physicians complained that Wang Xi had failed to even approximate Zhang Ji's original. When later Ming-Qing scholar-physicians reviewed the texts, however, they contended that the earlier scholars had not based their analysis on the authentic version of the *Treatise on Cold Damage Disorders*, which had been lost.[44]

The diseases of the south, which were more virulent than those in the north, led to new medical frameworks in Jin, Yuan, and Ming China. Epidemics challenged confidence in standard approaches to acute fevers derived from the canonical *Treatise*. Increasingly, physicians also questioned the government's formularies based on the Song medical orthodoxy. In addition, they historicized the concept of illness and interpreted the changing clinical landscape in terms of illnesses such as smallpox, which they saw as evidence of long-term changes in diseases brought by the winds.[45]

To understand the seasonal variability of illnesses in South China, and to grasp the role of warming or cooling as effective treatments, Song-Yuan-Ming physicians moved away from one-sided cooling or warming therapeutic strategies toward a more flexible approach that drew from both traditions. The heat-factor approach to the treatment of the epidemic diseases of south China became more accepted during the late Ming.[46] High mortality rates—of up to 70 percent in late Ming epidemics in the Yangzi delta, for example—led physicians to believe not that the medical classics were misguided, but that important parts of these works, especially the original *Treatise on Cold Damage Disorders*, had been improperly adapted.[47]

Ancients versus Moderns

Qing scholar-physicians sought to reverse the adulteration of ancient medical practice. Their appeal to the ancient wisdom in the authentic medical classics added to the growing eighteenth-century denunciations of the current medical orthodoxy. Moreover, Ming physicians such as Wang Ji (1463–1539) and Yu Chang (fl. 1630s) increasingly referred to case histories instead of the medical classics to advertise their therapeutic successes and explain them to students and amateurs.[48]

Qing debates between antiquarians and modernists concerning early medicine paralleled those between classical scholars. Like their counterparts in the classical fields, for example, Qing scholar-physicians began their studies with Han dynasty medical texts and the earliest classical interpretations, because these interpretations were written around the time of the texts themselves and thus were more likely to reveal their authentic meaning. Conversely, they rejected Song dynasty sources, on which "Song Learning" scholar-physicians had relied, because of their questionable authority: they had been written much later than the ancient texts.

Such approaches to the medical classics also led to investigations of the original content of the *Treatise on Cold Damage Disorders.* Editing and collating the variants for the current editions of the ancient medical texts enabled scholar-physicians to reexamine the original import of the medical classics. New works appeared on the *Inner Canon* and Zhang Ji's *Treatise*, and a proliferation of new annotations emerged.[49] Qing medical scholars demonstrated that Tang and Song medical works depended on Later Han texts that later interpreters had misread. The tense juxtaposition of an admired antiquity with a discredited Song medical orthodoxy suggests that medical studies in late imperial China were an adaptation of classical antiquity, one designed to improve contemporary medical therapies.[50]

In his exposition of Zhang Ji's *Treatise*, printed in 1648, Yu Chang reconstituted what he considered to be the 397 prescriptions of the original version.[51] Among the leaders of the Qing revaluation of the medical classics, too, Xu Dachun (1693–1771) advocated returning to the early medical classics such as the *Inner Canon* and the *Treatise*.[52] But Qing literati-physicians stepped beyond a mere reconstitution of the ancient text when they adapted the more contemporary heat-factor therapy to the cold-damage tradition, because they felt the two modes of analysis would provide more relief to patients when used together.

Qing followers of Jin-Yuan medical traditions remained prominent, and they opposed the efforts to reformulate ancient medicine. Called modernists by the editors of the Qianlong Imperial Library, these Jin-Yuan enthusiasts were opposed by Qing scholar-physicians because they relied minimally on the medical classics. Indeed, in 1739 the Qianlong emperor had already authorized compilation of annotations of Zhang Ji's *Treatise* in southern medical editions by Yu Chang and others. Published in 1743, it became the standard textbook for students in the Palace Medical Service. The Imperial Library catalog considered it evidence of the reconstitution in the eighteenth century of ancient meanings from the orthodox school of medicine.[53]

The shift from a universal medical doctrine (based on orthodox cold-damage therapy) to regional medical traditions (dealing with heat-factor epidemic diseases) began in the seventeenth and eighteenth centuries. Not until the late Qing, however, did the battle over the ancients and moderns end. By then, a collision with modern Euro-American medicine was imminent.[54]

Revival of Ancient Chinese Mathematics

Under the Ming, the mathematical traditions associated with the influential classic *Computational Methods in Nine Chapters* were continued. But most Ming literati no longer understood the pioneering methods for solving polynomial equations developed by the Song mathematicians. Consequently, in the eighteenth century, classical scholars increasingly focused on the work *Sea Mirror of Circular Measurement.* Written in 1248 by the minor official Li Ye (1192–1279), it was the oldest extant work on the "single unknown" techniques, which were used to solve polynomial equations containing a single variable.[55]

Recovery of Ancient Chinese Mathematical Works

During the Kangxi revival of interest in mathematics, Mei Wending and others could not easily find the works originally included in the medieval Ten Mathematical Classics. Moreover, in addition to Li Ye's *Sea Mirror of Circular Measurement,* the seminal works of Qin Jiushao (1202–1261) on polynomial algebra and other important topics were presumed lost. Remarkably, a large-scale effort to recover and collate the treasures of ancient Chinese mathematics became an important part of the surge in evidential studies during the late eighteenth and early nineteenth centuries.

In addition to famous evidential scholars who stressed mathematics in their research, a number of other mathematicians who were active in evidential studies edited ancient mathematical texts and digested European mathematical knowledge. They collated many of the mathematical texts under imperial auspices during the last years of the Kangxi reign, when the massive encyclopedia *Synthesis of Books and Illustrations Past and Present* was completed. When published in 1726, the *Synthesis* included some European texts from the late Ming and early Qing dynasties. Five works on Chinese mathematics were also added to the astronomy section of the encyclopedia.[56]

When the first set of the Qianlong Imperial Library collection was completed between 1773 and 1781, several of its compilers were well versed in classical mathematics. The Astronomy and Mathematics category incorporated fifty-eight works into the collection. Several older, lost mathematical texts were copied from the early Ming *Great Compendium of the Yongle Reign* (1402–1425), which had survived in the Imperial Court relatively intact. The general catalog of the Imperial Library, for example, included twenty-five notices on mathematics. Of these, nine were on Tang classics; three were on Song-Yuan works; four were on works from the Ming period, including the Ricci-Xu translation of Euclid's *Elements of Geometry;* and nine were on works from the Qing, including most importantly the *Collected Basic Principles of Mathematics* compiled with Jesuit help, and several works by Mei Wending.[57]

The recovery of ancient mathematical works occurred beyond the borders of the Qing dynasty as well. In particular, Korea and Japan played an important role in preserving lost Chinese works. For example, Ruan Yuan (1764–1849) recovered a lost mathematics primer by Zhu Shijie (fl. end of the thirteenth century) from a 1660 Korean edition that dated back to 1433. When transmitted to Japan, Zhu's primer and its single unknown algebraic notations helped guide advances in mathematics in Japan in the seventeenth century. Some have claimed, too, that a nineteenth-century Chinese work on the accumulation of discrete piles as a finite series—a cutting-edge mathematical principle of the day—was inspired by Seki Takakazu's (1642?–1708) *Compendium of Mathematical Methods* published in 1712.[58]

After the famous Korean calligrapher Kim Chŏng-hui (1786–1856) and several Korean emissaries visited Beijing and met Ruan in 1810, Kim sent Ruan the Korean edition of Zhu's primer, and Ruan presented a number of his works to Kim. At the time, Ruan and others were especially interested in Zhu Shijie's role in the formation of single unknown methods, which they explained in an

1839 publication. The Jiangnan Arsenal's Translation Department reprinted it in Shanghai in 1871 to show its relevance for mastering modern algebra.

While the Ten Mathematical Classics were being partly reconstituted, the Song-Yuan works of Qin Jiushao, Zhu Shijie, and Li Ye, among others, also became available. A special edition of seven of the Ten Mathematical Classics was reprinted by the Imperial Printing Office, and traditional mathematical works were also reprinted in several important collectanea.[59]

Reconstruction of the Ten Mathematical Classics

During the late Ming, Xu Guangqi claimed that the Mathematical Classics, which had provided a set of classical problems to solve, were inferior to Jesuit mathematics. Twelve works originally had been used to teach mathematics during the Sui (581–618) and Tang dynasties at the Imperial Academy (Appendix 1).[60]

No complete collection of the Ten Mathematical Classics was created until the eighteenth century, however, and only the *Gnomon of the Zhou Dynasty and Classic of Computations* was widely available in Ming times. Extant portions of the Ten Mathematical Classics in the late Ming were derived from Southern Song editions or from the early Ming *Great Compendium.* Not until the 1728 publication of mathematical texts in the *Synthesis of Books and Illustrations* encyclopedia did the work of collating the Ten Mathematical Classics begin in earnest. The celebrity that Mei Wending had achieved as a mathematician, coupled with the publication of several new European mathematical works during the late Kangxi reign, brought mathematical astronomy into the mainstream of classical studies.

Indeed, the influence of the *Collected Basic Principles of Mathematics,* as a translation of what the Qing court thought were the latest features of European mathematics, stimulated scholars associated with evidential studies to rediscover the Chinese origins of Western mathematics. While serving on the Imperial Library staff in the 1770s, for example, Dai Zhen collated seven of the ten mathematics classics from the *Great Compendium of the Yongle Era.* In addition, he recovered two more from manuscript copies originally held by the Mao publishing family. The Imperial Printing Office published these as rare editions in a special collectanea. Dai's colleague Kong Jihan (1739–1784) then had them reprinted in 1773 under the title "Ten Mathematical Classics," which represented the first collectanea with this name. Subsequent editions, including those with Dai's annotations, were based on these versions.[61]

Rediscovery of Song-Yuan Mathematical Works

Reconstructions of the "single unknown" and "four unknowns" techniques for solving polynomial equations that featured several unknown variables and included variables with exponents were particularly prominent in the late Qianlong era. Qin Jiushao's *Computational Techniques in Nine Chapters* (1247), for example, provided general algorithms for solving the Chinese remainder problem. He also investigated techniques similar to the Horner-Ruffini method devised in the early nineteenth century for calculating the roots of polynomial equations.[62]

Although Qin's work followed the overall structure of the *Nine Chapters,* his algorithms were much more sophisticated. In addition, because Qin had studied in the Song Astrocalendrical Bureau as a youth, his work treated certain astronomical predictions as problems in remainder theory. Finally, the *Nine Chapters* calculated the area of any triangle as a function of the lengths of its three sides, which was similar to the prototrigonometric relational features method for computing the sides of a right triangle revived in the late Ming. Later, this approach drew the attention of Dai Zhen and others interested in correlating such relational features with Jesuit trigonometry.[63]

After the Mongols conquered North China in 1232, Li Ye prepared the *Sea Mirror of Circular Measurement* (1248), which was copied into the eighteenth-century Imperial Library. Although Li lived in reclusion after the Mongol triumph, he was called to the Mongol court to consult on governance and earthquakes. The *Sea Mirror* survived in a book in Li Huang's private library. It presented 170 problems on single unknown techniques. Li based the problems in it on a single diagram, a triangle with an inscribed circle representing a city wall (Figure 2.1). Using a method equivalent to polynomial equations (including negative powers of an unknown), Li calculated the diameter of the circle and the lengths of segments in the figure.[64]

The mathematical works of Yang Hui (fl. in Hangzhou during the Southern Song) were all included in the *Compendium of the Yongle Reign,* although they were not copied into the Imperial Library. Some of Yang's works were edited and printed in Hangzhou by Qing bookmen. In 1840, Yu Songnian included Yang's commentary and supplement for the *Nine Chapters* and *Yang Hui's Calculation Methods* (1275), which was still incomplete, in a collectanea. Yang's complete work was lost in China, but it was rediscovered in the late Qing. In the early twentieth century, too, a 1433 Korean edition of the *Calculation Methods,* based on a 1378 Ming edition, was located. It was recopied in Japan by Seki Takakazu when Japanese scholars discovered it among the books that had been brought from Korea by Hideyoshi's retreating forces.[65]

Other works, such as Zhu Shijie's *Jade Mirror of the Four Unknowns* (1303) and his *Primer of Mathematics* (1299), were not recovered in time for inclusion in the Qianlong Imperial Library. Although Zhu's *Jade Mirror* purported to solve practical issues dealing with architecture, finance, and military logistics, it energized late

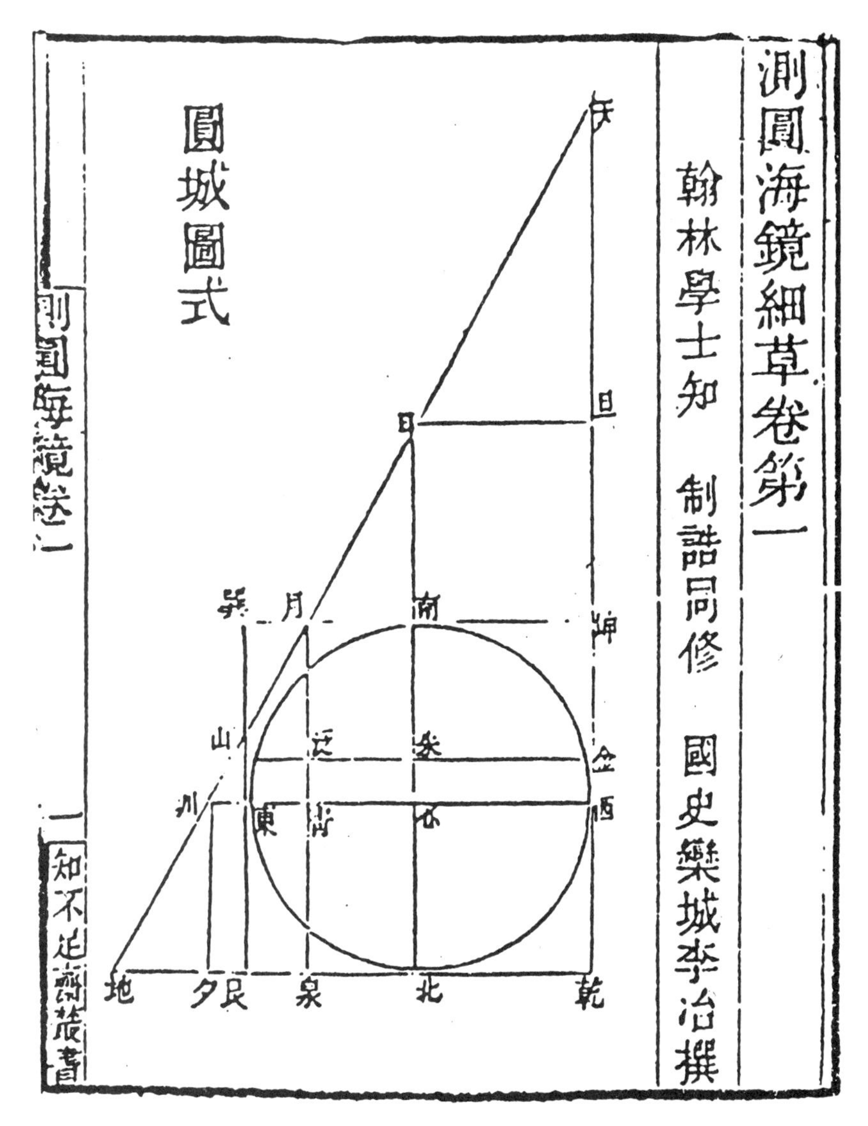

Figure 2.1. Li Ye's "diagram of a triangle with an inscribed circle" (representing a triangle around a city wall). *Source: Zhi buzu zhai* edition of the *Ceyuan haijing* (*Zhu buzu zhai congshu,* 1776–1823).

Qing evidential scholars who found in it a Chinese algebra to extract roots using counting rods that predated the Jesuit's borrowing roots approach. Zhu's polynomial equations went beyond the second and third degrees, up to the fourteenth.[66] While governor of Zhejiang, Ruan Yuan acquired from a Korean envoy his version of the *Primer of Mathematics,* which he used to reconstitute the *Jade Mirror of the Four Unknowns.*[67]

The discovery and dissemination of these ancient works provided important clues about the fundamentals of Song-Yuan polynomial algebra.[68] Moreover, they were an important part of a contemporary interest in antiquarianism, which was fueling the inward turn to native mathematics. Chinese classical scholars at the cutting edge of evidential studies grasped the importance of advanced algebraic techniques for solving complicated equations based on sophisticated mathematical problems. At the same time, however, they were highly focused on the recovery of ancient texts. The combination of interests led to a high level of mathematical skill among some Chinese. When Protestants finally introduced the differential and integral calculus in the middle of the nineteenth century, for example, Li Shanlan (1811–1882) and others appreciated its sophistication because they had already mastered the skills needed to solve "single unknown" and "four unknowns" problems.[69]

Ruan Yuan and the Biographies of Mathematical Astronomers

Ruan Yuan's compilation of the *Biographies of Astronomers and Mathematicians* while he was governor of Zhejiang from 1797 to 1799 climaxed the celebration of natural studies within the Yangzi delta literati world. *Biographies* contained summaries of the works of 280 mathematicians and astronomers, including thirty-seven Europeans, and was followed by four supplements in the nineteenth century. The collection was reissued in 1829 with only Qing biographies, and it was later enlarged and reprinted. In 1840, for example, Luo Shilin (1789–1853) added forty-three sections on Song-Qing mathematical astronomers based on new sources for the four unknowns techniques recovered in 1822 from the Song and Yuan. And in 1857, Alexander Wylie (1815–1887) worked with Wang Tao (1828–1897) to improve on the views presented in the collection.[70]

Ruan was aided in his Hangzhou *Biographies* project by leading evidential scholars of the late Qianlong period. Ruan did not include fortune-telling or numerology in the collection, and he opposed connecting mathematical astronomy with mathematical harmonics or studies of the *Change Classic.* He did address, but was critical of, three new findings introduced by Michel Benoist (1715–1774): (1) heaven and earth are round, (2) planets follow elliptical paths, and (3) the sun is stationary. Although Benoist had by then finally presented the Copernican system to China as the European norm, Ruan Yuan found such views unacceptable, in part because they contradicted earlier Jesuit presentations of Copernicus, which had considered his cosmology to be an offshoot of the Tychonic system. Ruan sought a fusion of European and Chinese mathematics based on shared common conceptions. For astronomy, he sought an accurate, predictive computational system that would be based on improved techniques, not cosmology.[71]

Interest in mathematical astronomy among the Chinese literati was tied to the popularity of evidential studies outside the patronage networks of the Manchu court. By connecting mathematics and astronomy to classical studies, Ruan Yuan successfully integrated mathematical astronomy with evidential studies. He and his colleagues also revived the ancient term "metrologists" (*chouren,* that is, those associated with computational astronomy, mensuration, and surveying) for those whom he now considered mathematical astronomers.

In the mid-eighteenth century, the official *Ming History* recounted the ancient dispersion of classical metrologists to the "Western region," specifically to the Islamic world. Moreover, the term for classical metrologist had been used in the Han dynasty canonical writings. Such usage was now reworked by Ruan Yuan in his lead account of the meaning and scope of mathematical astronomers in the *Biographies.* He employed the term as a classical sanction for a new intellectual and social category of contemporary scholar-literati such as Mei Wending, implying a genealogy of professionalized skills in mathematics and astronomy going back to antiquity. This orthodox term was the first of several used in the eighteenth and nineteenth centuries to describe European scientists using classical Chinese.[72]

Beginning in the late Ming, as literati increasingly engaged in the study of mathematics and astronomy, two types of experts emerged: computational astronomers, and literati with an academic interest in mathematics. These two categories were most evident during the Kangxi era, when mathematics was needed for calendar reform. The academic climate among evidential scholars, along with imperial patronage, helped make mathematics and astronomy a collateral branch of classical learning.[73]

The Qing literati justified natural studies as the proper concern of the scholar-official precisely because scholar-officials were included in the classical system. Experts, as long as they were subordinate to dynastic orthodoxy and its official representatives, were necessary parts of the cultural, political, and social hierarchies. The literatus-official coexisted with the astronomer, imperial physician, or Jesuit artisan in the bureaucratic apparatus, but was given higher levels of political status, cultural prominence, and social prestige.

Qing dynasty literati were increasingly conversant with mathematics before the First Opium War. Moreover, due to their mastery of Jesuit algebra and native techniques, they generally appreciated both.[74] Literati mathematicians were still few, however, and they lacked a Newtonian mechanics to find practical applications outside the domains of astronomy and cartography. Alexander Wylie's and Li Shanlan's 1859 preface to *Step by Step in Algebra and the Dif-*

ferential and Integral Calculus, which they translated from Elias Loomis's (1811–1889) *Elements of Analytical Geometry and of the Differential and Integral Calculus* (1851), noted that such reasoning on trigonometric series was close to the concept of infinitesimals and represented the starting point for studies of trigonometric expansions in the nineteenth century.[75]

Perhaps the most influential mathematician in China until his death was Li Rui (1768?–1817?), a disciple of Qian Daxin (1728–1804) at the Ziyang Academy in Suzhou. The complete works of Li Rui that Ruan Yuan published in 1823 included eleven mathematical treatises, with seven on ancient astronomical systems and four on equations and traditional methods for computing the sides of a right triangle. Ruan Yuan's account in the *Biographies of Mathematical Astronomers* had noted that single unknown procedures had fallen into disuse in Ming times. As a corrective, Li Rui worked out problems using single unknown procedures rediscovered in the eighteenth century, thereby restoring the mathematical saliency and relevance of pre-Ming mathematical texts.[76]

THE RISE OF IMPERIAL CHINESE MANUFACTURING AND TRADE

Before 1800 and the Industrial Revolution in Europe, Jesuit-supervised imperial factories in China produced many luxury arts and crafts. China's own immense porcelain industry, to choose but one example, produced millions of pieces for sale in the eighteenth century, which suggests an interesting parallel to the rise of Wedgwood porcelain and mass production in England during the Industrial Revolution. But it is puzzling that the transition in Qing China from imperial-style factories to modern industry did not occur until the late nineteenth century. If the Chinese were relatively open to Western learning in the eighteenth century, why the wait? There were many external factors that explain why the Newtonian revolution reached China very late.

Accounts of modern science in the eighteenth century often contrast the French sciences with practical engineering in England, even though recent work points to the overlapping cultural values that fostered disciplined curiosity among natural philosophers, capitalists, entrepreneurs, educators, engineers, and industrialists in both countries. British and French citizens, the successors to the Jesuits, introduced what they thought of as their own regime for science to China in the nineteenth century, but in fact here too their agendas had some commonalities.

The Jesuit Role in Qing Arts, Instruments, and Technology

Before the Napoleonic Wars were resolved in Britain's favor—that is, before 1815—the Qing court and its literati continued to co-opt certain aspects of European learning, artisanal know-how, and artistic expertise. For example, Jesuits remained as court clockmakers and geometers, and a Jesuit, usually Swiss, was always in charge of the imperial clock collection in the eighteenth century. Yet when compared to the industrializing tendencies in Britain and France, the imperial workshops in mid-Qing China, like the Academy of Mathematics under the Kangxi emperor, never featured cutting-edge European science.

Qing artisans—usually Manchu bannermen and Chinese bond-servants serving in the imperial workshops—were also important in manufacturing clocks, porcelain, glass, and building pavilions and gardens during the late Qing empire. Interestingly, when Great Britain sent Lord Macartney to China in 1793 to try to open diplomatic relations, he mimicked the Jesuits and presented clocks and watches, as well as a telescope and planetarium, to the court to demonstrate European inventiveness. But Qianlong already had such manufactures and was not particularly impressed. He rejected any official trade with Britain and with his haughty imperial tone belied that he extensively collected and reproduced European manufactures and clockwork at home.[1]

Clock Making in the Kangxi Era

European expertise in mathematics, astronomy, cartography, and clock making was connected to the unprecedented degree of accuracy in European time and space measurements, which by 1750 were required as precise entries in the engineering equations that measured velocity, acceleration, and force. But mechanical clocks in early modern Europe were also a metaphor for God's maintenance of the universe. In an age when a mechanistic view of nature emerged, Europeans believed that the mastery of clock mechanics would lead to a better understanding of God's design of the world. The Jesuits accepted this view, but they also recognized that when they presented mechanical clocks to potentates at home and abroad, they gained access to high places in the papal court.

The Jesuits' skills as architects and glass makers paralleled their careers as computists and clock makers in late imperial China. Matteo Ricci believed, for example, that he had gained admission to the Ming court by presenting a clock and repeating watch to the emperor, a tactic that followed established European gift-giving practices. When Johann Schreck (also known as Johann Terrenz) and Wang Zheng prepared their 1627 translation entitled *Diagrams and Explanations of the Marvelous Devices of the Far West,* it represented the first work in Chinese that provided information about European escapement techniques for delivering power at regular intervals to move the gear train inside a timepiece. Subsequently, the Ming Wanli emperor (r. 1573–1619) had his eunuchs work with the Jesuits to master the art of repairing the clocks.

In the late seventeenth century, the Kangxi emperor established a number of workshops, modeled after studios in the French Academy, that were to manufacture luxury items under the auspices of the Office of Manufacture in the Imperial Household Department. Chinese clockmakers were trained as well; the Kangxi emperor thought that modern clocks were an improvement over the traditional water clocks then in use, and the Chinese who worked under Jesuits were expected eventually to make their own mechanical clocks.[2]

The Kangxi workshops, then, were staffed by Jesuits and Guangdong workmen, and after they were formally established in 1693, there may have been as many as thirty-one such shops. Most clocks in the eighteenth century were completed in shops inside the Forbidden City. A specific Office of Clock Manufacture was subsequently created in 1723, which lasted until 1879, when the last list of court clocks was compiled.[3]

Three major manufacturing sites were operating in 1756 within the Imperial Household Department. There clockmakers, painters, and engravers worked in the decorative arts using enamels and glass for the clocks. The Lofty Pavilion Garden-Palace, one of these sites, also served as a venue for clock making. In 1752, Jean-Joseph-Marie Amiot (1718–1793) mentioned that the Jesuits had prepared for the empress dowager's sixtieth birthday a mechanical theater clock with three scenes, which remained in the Lofty Pavilion. Manchu and Chinese interest in elaborate mechanical clockwork, much like that in Europe, reflected connections to imperial power and prestige.[4]

Five European horologists were working in Beijing in 1701 during the Kangxi reign. Under the Qianlong emperor, eleven Jesuits built clocks in the imperial workshops. By 1800, with only a few ex-Jesuits remaining, Chinese artisans were proficient enough to make mechanical clocks themselves, so the missionaries became less involved in production. Outside of Beijing, for instance, imperial clock-making workshops were also established in Suzhou and Hangzhou.

The native industry for clocks provided a commodity that served as an effective form of tribute among Qing officials. Native artisans, then, who made very few items compared to the massive porcelain industry (and whose work was controlled by Chinese merchant guilds), looked to the court and its literati elites as consumers. Indeed, mechanical watches were very popular among eighteenth-century Chinese elites. Mechanical clocks were mentioned in novels such as *Dream of the Red Chamber,* in which owning one was described as reinforcing the high status of a Chinese family.[5]

The Qing court and its Jesuit clockmakers did not realize, however, that more than just connoisseurship was involved in the development of more accurate timepieces in Europe. Artisans and engineers there were urgently demanding more accurate time measurements, so as to solve practical equations analyzing the volume flow of liquids or the velocity of solid objects. In addition, seafaring captains and their officers needed to establish the exact latitude and longitude of their ships at sea. What drove precision clock making in the British navy, for example, was the recognition among captains that small errors in navigation aboard their vessels could prove fatal if the ships ran aground—and that the alternative, traveling only well-traveled paths on the ocean, made their ships more vulnerable to pirates. By 1714, the British Parliament was offering a sizeable prize to scientists who could "determine longitude to an accuracy of half a degree." Subsequently, accurate and synchronized timepieces were created that when on board could determine a ship's distance from a standard zero-degree meridian.[6]

Unlike Qing China, which prided itself as the "Central Kingdom" with Beijing as its continental hub, Greenwich, England, became the center of the British navy and its efforts to measure and synchronize a ship's distances globally. The cutting-edge timepiece on Captain James Cook's triumphal second voyage of 1772, for instance, was not an idle curiosity (as clocks were in China), but a byproduct of Newtonian mechanics and the calculus, which had linked time and space to map movement. It was the key to plotting accurately the location of his ship throughout the voyage. The British East India Company soon discovered another important byproduct of accurate timepieces: their agents abroad could adjudicate contested boundaries in their favor using impartial longitude measurements based on time and speed traveled.[7]

Imperial Factories for Glassware

The French Jesuits remained involved in the glass workshop that Kangxi established in 1696 in the palace, and the German Johann Kilian Stumpf worked with them to produce decorative glass under imperial auspices. To please the emperor, the Jesuits arranged to have two European glassworkers arrive in Beijing in 1699 to work in the imperial workshop. As a result of these efforts, the Beijing workshop was producing high-quality glassware from the beginning of the eighteenth century. Decorative snuff bottles were one item produced. Fabricated in a wide variety of colors and shapes, such glassware was presented as gifts by the emperor to high officials on the emperor's tour of the Yangzi delta in 1705. He also presented papal legates with an enameled glass snuff bottle at the outset of meetings in 1706 to discuss the Rites Controversy.[8]

Matteo Ripa (1682–1745), an Italian Jesuit who arrived in China in 1710, noted that by 1715 the glass workshop consisted of several furnaces for glassmaking, a task that required a large number of skilled craftsmen under Stumpf's supervision. After a French Jesuit expert in enamel arrived in Beijing in 1719, the enameled glassware produced in the workshop became of high enough quality that in 1721 the Kangxi emperor sent the pope two large cases of enamelware, in addition to 136 pieces of Beijing glass.[9]

After Qing anti-Jesuit policies took effect, and particularly after Stumpf died in 1720, the imperial workshop, like the Academy of Mathematics, increasingly relied on native glassmakers. The court also encouraged the workshop to manufacture enamel colors independently: it sent enamel colors to the imperial pottery kilns in Jingdezhen, where new, enamel-decorated porcelains appeared for the first time.[10]

Despite his animosity toward the Jesuits, the Yongzheng emperor established a branch of the glass workshop in the Lofty Pavilion Garden-Palace in the northern suburbs of Beijing. Production also continued at the Jesuit glass workshop into the Qianlong era, when two additional Jesuits joined the imperial glass workshops in 1740.

Glass production reached its high point in the 1750s when, at the request of the emperor, the Jesuits became involved in decorating European-style palaces and gardens that were being constructed for the Lofty Pavilion by Giuseppe Castiglione (1688–1776) and others. Chinese decorative themes were influenced by European skills in shading and adding perspective to illustrations on glassware. By 1766, the Qianlong emperor had turned the Lofty Pavilion gardens into a treasure house of gardens, pavilions, paintings, glassware, porcelains, and furniture, whose artistic merits and technical prowess were of the highest standards.[11]

Jesuits and Garden Architecture

The Jesuits found new life in the Qing cultural world of the 1750s through their expertise in painting, designing, and building the lavish European-style palaces and interiors that Qing emperors sought. Impressed with the grandeur of European-style fountains, Qianlong asked Castiglione to join his imperial design offices to draw up plans for such fountains in the Lofty Pavilion gardens. Castiglione in turn sought the help of Michel Benoist, who had previously presented the court with an accurate account of Copernican cosmology. Because Benoist was knowledgeable in mathematics and hydraulics, as many Jesuits were, he was able to present the emperor with a model fountain, which the court quickly authorized Castiglione to build alongside the Baroque-style palatial buildings.[12]

The European-style garden designers were Jesuits, with Castiglione playing the most prominent role from 1745 to 1759. They were well prepared for their task in Beijing because in contemporary Rome the popes had glorified themselves and the Church with lavish palaces, piazzas, and fountains whose construction had been supervised by the clergy throughout the old city. During the mid-eighteenth century, European rulers and aristocrats were avid connoisseurs of chinoiserie architecture, gardens, and lakes, a trend that led to the construction in Europe of Chinese-style rooms, gardens, and pagodas.[13]

Overall, the Lofty Pavilion represented an eclectic architectural style that the Manchu court favored as part of their efforts to create a universal vision of their power in Asia and beyond. The interior designs for the pavilions were inspired by European models with glass windows, wood plank floors, handrails, flower terraces overlooking lawns, mechanical clocks, hanging lamps, and oil paintings. Even Gobelin tapestries of French beauties presented by Louis XV were placed throughout the European section. Chinese garden elements outside and traditional scroll paintings inside added to Manchu pretensions as global style setters.[14]

Porcelain Factories in China

Until 1600, Ming artisans manufactured stoneware (fired at 1,100 to 1,200 degrees centigrade) and porcelain (fired above 1,300 degrees) primarily for the huge domestic market of some 150 million people, principally at local coal-firing kilns in the north and wood-firing kilns in South China. Merchants passed the high-fired stoneware and ceramics to their imperial and literati consumers, while traders also provided information to the local producers about market demand.

Via lakes and rivers, the highest quality porcelain in South China also found its way to southeastern ports and from there to the Indian Ocean trade and Islamic markets. And when the owners of the southern kilns lost the patronage of the financially strapped late Ming court, they exported porcelain to Japan and Southeast and South Asia instead. The Dutch East India Company alone handled some six million "blanc-de-chine" pieces made at Dehua, Fujian, in the seventeenth century, but this number represented only about 16 percent of Ming ceramic exports.[15]

Jingdezhen Porcelain

Dating from 1004–1007, the largest pottery factories were in the southwestern province of Jiangxi at Jingdezhen, where wood was plentiful, the clays were ideal, and labor was relatively cheap. Jingdezhen followed the usual imperial pattern for operating such enterprises, which involved state supervision of merchant activities. The factories relied on the labor of thousands of artisans to produce the high-quality celadon and blue-white high-fired wares for which the Ming became famous. Imperial taste and literati connoisseurship deflected the technical discourses of the producers into a sublime discussion of porcelain aesthetics for highbrow consumers (Figure 3.1).[16]

When the Ming fell in 1644, imperial patronage ended, but Jingdezhen revived and remained the major domestic and international producer of porcelain in China after the Kangxi emperor ordered the kilns rebuilt in 1682 or 1683. Manchu emperors quickly imposed their imperial tastes on the porcelain made during their reigns. The Kangxi emperor, for example, favored rich sapphire-colored pieces fired with local cobalt oxides, which replaced the "Mohammedan blue" fired using Persian cobalt during the Ming.[17]

In early modern Europe, Chinese porcelain was called "china," and it was very widely known and appreciated. For example, August the Strong (1670–1733), king of Saxony, was so enamored of the style that he had over twelve hundred blanc-de-chine wares, and founded the Meissen porcelain factory. Until their popularity waned in the eighteenth century, blanc-de-chine pieces were also collected at the English royal palace at Hampton Court, and they appear in the 1688 catalog of holdings in the Cecil family's Burghley House. By 1700 more than 150,000 pieces of porcelain might be unloaded from an arriving ship in England, and the British East India Company took some 400,000 orders by 1722 to satisfy the obsession for "chinaware" among upper-class women.[18]

During the Yongzheng reign, the bondservant Nian Xiyao (1671–1738) became director of the Jingdezhen factory. He included European designs that had been introduced as result of Castiglione's influence at the Manchu court, which led to a synthe-

Figure 3.1. Blue and white porcelain bowl from the Qing dynasty. Shandong Provincial Museum, © 1984. China Institute in America.

sis of Chinese and European art. Nian had a broad background in the arts and sciences. Prior to becoming director, and while Jesuit influence on court mathematicians remained important, Nian had written several works on mathematics. In addition, the techniques of perspective in European painting, which he had learned from Castiglione, impressed him.[19] Perhaps his leadership caused in part the significant increase in mass production of porcelain at Jingdezhen that occurred particularly in the Qianlong period. Artisans during this time also applied Jesuit enamel-painting techniques as they decorated high-fired ceramics and cloisonné.[20]

Techniques for Porcelain Production

Developments at the Jingdezhen factory drew the most attention from Europeans, who sought the Chinese secret for making porcelain. Englishmen such as Francis Bacon (1561–1626) thought that porcelain was developed from artificial cement buried in the earth for a long time, whereas John Donne (1572–1631) thought buried clay was its source. In 1712 the French Jesuit Father Francis Xavier d'Entrecolles (1662–1741) visited Jingdezhen. His detailed account of the sloping dragon-shaped kilns used there since the Mongol era touched off a European search for comparable ingredients to manufacture porcelain at home.[21]

Although porcelain was relatively unknown in Europe until 1675, by 1715 it had become a significant feature in aristocratic households. Europeans were unsuccessful in making porcelain until the Meissen factory in Germany produced the first European hard-paste porcelain in Dresden in 1710. After d'Entrecolles's 1712 account, however, the famous porcelain maker Josiah Wedgwood (1730–1795) used the Jesuit priest's descriptions to help him create the floor production plan for his first English factory. In 1759 this factory began producing low-fired cream-colored earthenware.[22]

Artisans began by mixing "kaolin," a fine white refractory clay, with *petuntse* (literally, "small white briquettes") from a rock known as pegmatite. Jiangxi quarries outside of Jingdezhen provided the kaolin. Hydraulic hammering devices crushed the clay chips, and workers placed the white powder into water where it was stirred and decanted. When a solution of fine particles rose to the top, it was repeatedly collected until at the bottom there remained only paste, which was then transferred to large molds and allowed to harden. After the powder was transformed into soft paste, it was purified.

The workers delivered the lumps of clay to the potter's workshop for fashioning and shaping on the wheel. After the piece dried in the open, the potter finished it with a turning tool, carving or incising designs. Another craftsman prepared a mold out of yellow clay for other required shapes (for complicated pieces, several sections might be required for the mold). After drying, potters coated the pieces to prevent the coloring oxides from being absorbed during firing.[23]

The mixture fused when artisans heated it to a temperature of 1,280 to 1,350 degrees centigrade, creating genuine hard-paste porcelain. For blue hues, they diluted cobalt oxide in water and painted it on before applying the outer glaze for decoration. They then applied a final coating of glaze until the piece was white. During the enameling process, potters applied Jesuit-inspired rose hues as a decorative underglaze. They then applied another coating of *petuntse* mixed with gypsum, fern ash, and quicklime for a second firing at lower temperatures of 600 to 800 degrees centigrade. Potters could also stain the piece with a metallic oxide or ash to achieve a wider range of hues.

Workers placed the pieces to be fired inside the kiln in containers called saggers, which were made of refractory earth or porous clay. They covered each piece with a layer of sand and powdered kaolin to keep the piece from sticking to the container. More delicate porcelain was placed in the middle of kiln. Smaller pieces could be fired for five days. Larger pieces might take up to nineteen days, which included several firings (and cooling periods) to bring out the decoration.

Workers carved several spy holes along the sloping kilns to enable them to note the melting markers, which told them when the kiln had reached the required temperature. Once the coating of glaze had vitrified onto the body of the piece and cooled, the potter applied an overglaze to the surface for additional firings. Depending on the glaze, this process might produce willow-green celadon; shiny blue or mirror black, which was popular for wares of the Kangxi period; or the yellow, rust, and "tea-dust" colors so popular for wares of the Qianlong period. Copper oxide produced red and green, as well as the peach-bloom and turquoise-blue glazes.

Upon completion, workers opened up the kilns in the presence of an official, who took 20 percent of the contents as tribute for the

Qing dynasty. Potters completed the touching up at the workshop before they sorted the pieces according to quality. As many as 50 percent might be rejected. They sold a certain number of pieces immediately, while others were transported in wooden casks from the remote area of Jingdezhen through the river and lake systems of the middle Yangzi region to major urban centers and then on to Beijing.

Guangzhou became the gateway for the porcelain trade to Southeast Asia and Europe. Aeneas Anderson, who accompanied Lord Macartney on his 1793 mission, wrote: "There are no porcelain shops in the entire world which can compare in size, richness or delivery with those in Canton." After 1750, European chemists discovered that the technical secret for making ceramic ware occurred during firing: the combining under very high temperatures of infusible materials (such as kaolin) and fusible materials (such as *petuntse*) resulted in the glass-like mixture called porcelain. As a result of this technology transfer to Europe, as well as the drop in Chinese imperial patronage during the nineteenth century, the Chinese porcelain industry lost its standing among foreign connoisseurs and declined significantly in the international market. Wedgwood in England and Japanese potters replaced Jingdezhen as the best quality kilns in the world and gained the upper hand in world trade. Meanwhile Jingdezhen and other regional kilns continued to supply the huge Chinese market with stoneware and porcelain.[24]

Printing Technology and Book Publishing

Woodblock printing had already developed during the Song and Ming periods into a sophisticated art. Indeed, classical learning revived during the Song dynasties in part due to the increased circulation of books and the diffusion of classical texts brought on by the spread of printing. Government printing during the Song had also encouraged paper production (invented in the second century

C.E.). The existence of a large, empirewide paper industry was in place to support the large-scale publication of books, and by the late Ming, multicolor printing, artistic illustration, copper movable type, and woodcut facsimiles of earlier editions were all being used.[25]

The explosive expansion of printing after 1500 yielded a wider circulation of erudite and practical knowledge empirewide. By 1800, scholarship, book production, and libraries were central to the Chinese culture. While literati elites in the southern areas embraced classical works, a broader audience located in the Yangzi delta, as well as in southeast and southwest China, welcomed the printing of popular literature, vernacular novels, almanacs, encyclopedias, and literacy primers.[26]

Woodblock printing reached its peak of technical sophistication in the mid-sixteenth century with the rise of scholar-printers in the Yangzi delta and more commercially oriented printers in Fujian. During the late Ming, Nanjing, the Ming southern capital, and nearby Suzhou (both cities are in the Yangzi delta) became the center for quality printing, and outstanding xylographers staffed the printing shops. Jianyang (in Fujian), the center for commercial publishing, produced the most novels, dramas, and popular manuals (including medical handbooks). Merchants from Huizhou (in Anhui), who could access the cheaper wood in their home area for woodblocks, as well as book traders from throughout China, congregated in Nanjing as well as Hangzhou, the former Song capital.[27]

Family businesses predominated in Fujian publishing, particularly where local, low-cost enterprises had been created through an intricate network of bookstores and traveling merchants. The kinship links among local publishers, distributors, and sellers supported a nested geographical hierarchy for producing and selling books for elite and popular audiences locally and regionally. As family businesses, the local print shops fulfilled family members' cultural aspirations by catering to the local civil examination market for books on the Classics and classical primers. In this manner, merchant families that profited from the book trade also invested in education for their kin in the hope that their social status would improve through an association with books and classical scholarship.[28]

Although Chinese printers experimented with movable type, xylography was generally—but not always—more economical, when publishing many copies of a particular work. In general, woodblocks were easily stored and, with reasonable care, could be preserved for frequent reuse. In fact, printers often used the same woodblocks for collectanea of different titles and compilers. Woodblocks were especially economical for books and manuals produced by lowbrow print shops in Fujian and Sichuan. By saving a woodblock, printers were spared the time-consuming (and thus costly) task of breaking down and recasting movable type every time a new printing of a popular book was required. Yet in some instances—such as when relatively few copies of a work were needed (say, for imperial projects with limited circulation)—movable type was the most economical option (Figure 3.2).[29]

The proliferation of books and manuals during the late Ming led to the printing of numerous encyclopedias and classified digests. These books functioned as repositories and manuals of popular knowledge during the late Ming, in addition to serving as scholarly compendia for students preparing for the civil examinations. The book-oriented atmosphere that emerged from this environment turned out to be very conducive to the development of scholarship and the practical arts.[30]

In addition, the printing of "daily use" compendia since the 1590s broadened the appeal of published works to the lesser lights of Ming-Qing society—namely, merchants, artisans, and those licensed only to take the lower examinations. These popular digests, which were presented as repositories of useful information for daily life, provided nonelites with a wide array of information on sub-

Figure 3.2. Setting movable type in the Qianlong Imperial Printing Office. *Source: Qinding Wuying dian juzhen ban chengshi* (Beijing, 1776).

jects such as medical prescriptions, divination formulas, ancient lore, astrology, geomantic almanacs, calligraphy, and so forth.[31]

Ribald novels such as *The Plum in the Golden Vase* were not only themselves examples of popular culture during the era; they also offered a view of contemporary lowbrow life, at least as described by the highbrow compendia and encyclopedias that were available to the authors. Circulating as a manuscript in the late 1590s, the *Plum* presented an inventory of things, money, objects, collectables, events, and skills that ranged from medical potions for enhancing sexual prowess to food offered at banquets, drinking games, and popular jokes. The fictional contents of the novel thus not only enlivened but also mirrored the categories and contents of the narrativeless encyclopedias.[32]

In addition to their book collecting efforts, bibliophiles in the Yangzi delta undertook major printing projects. Publishing had been an important way of building large private collections since family printing efforts began in the late Ming. The bookman Bao Tingbo (1728–1814), for example, used his Hangzhou collection to publish rare editions of mathematical texts and manuscripts in his possession under the collective title *Collectanea of the Can't Know Enough Studio* (1776). As works were added to the collection, the collectanea encompassed a total of thirty series. Scholar-printers like Bao became invaluable members of the Yangzi delta academic community.

The Book Trade in the South and Beijing

Rare books were already expensive during the Kangxi era, and by the Qianlong period their value had increased tenfold. Private libraries flourished in the social and academic environment of the Qianlong reign. During the early Qing, the bookstalls in Beijing, where new, old, and rare books and manuscripts were sold, were located outside the city at Ciren Monastery and frequented by scholars, who had lived in the area since the 1660s. Literati who were famous as painters or as connoisseurs of ancient paintings and artifacts were also frequently consulted on the purchase of rubbings and the like.

By the eighteenth century, the Glazed Tile and Glass Factory, located on a street in the southern, Han Chinese quarter of Beijing, became the major book emporium and center for antiques in Qing China. Originally a factory site, the Glazed Tile and Glass Factory reached its height as a book market during the Qianlong era. Because it was located just outside the Forbidden City, close to the Hanlin Academy (where the imperial scholars and Grand Secretaries were appointed), the emporium was a gathering spot for intellectuals, scholars, and civil examination candidates who came to Beijing. Its cultural atmosphere stressed the value of rare works and ancient artifacts. Moreover, vendors promoted the exchange of books and stimulated scholarship during the eighteenth century. Books and manuscripts of all kinds moved freely among Beijing and the markets in the Yangzi delta and southeast China.[33]

The book trade in China also attracted the interest of scholars from Korea, who accompanied their country's tribute missions to Beijing. Koreans noted that many of the merchants in the Glazed Tile and Glass Factory were from the south, which suggests the important role played by Yangzi delta proprietors in the book trade. Chinese booksellers, for instance, had close discussions with the Korean scholars regarding rare books and editions available in China and Korea. Pak Che-ga (1750–1805?), a leader in the eighteenth-century Korean classical learning, was one such scholar who was attracted to the Chinese academic environment and book trade. In his pursuit of information on Qing institutions, technology, administration, and evidential research, he visited Beijing four times beginning in 1776 and associated with leading classical scholars and men of letters.[34]

Korean scholars had been visiting the Glazed Tile and Glass Factory in Beijing since at least the Kangxi era, looking for books to take back to their homeland. In the seventeenth century, when Manchu restrictions made it impossible for Korean envoys to leave their compound, Korean scholars purchased books from merchants with licenses to trade with diplomatic missions in Beijing. A process of cultural exchange and correspondence ensued that closely linked eighteenth-century Korean learning to Qing classical scholarship. Imperial Library editors also developed a warm relationship with a number of the Korean scholars who accompanied the Korean tribute missions.

In search of books to add to the Korean royal library, which had been launched in 1776, Korean scholars, with the help of their Chinese friends, collected a number of series and encyclopedias. Korea contributed a large amount of epigraphic material to Qing scholars. In Chapter 2, I described how Kim Chŏng-hui sent back to Korea a rare Yuan dynasty mathematical text that had survived. For his efforts, Ruan Yuan sent Kim his influential Qing classical compendia as soon as it was published in 1829.[35]

After the Manchu conquest of China and Taiwan was made secure in the 1680s, the Chinese presence in trade between the ports of Ningbo and Nagasaki increased, with as many as ten thousand native Chinese living in the Chinese quarters in Nagasaki. Among the important commodities in that trade were recent books published in China. Because of the Tokugawa ban on Jesuits in Japan after 1637, Japanese scholars and shoguns especially sought Chinese translations of European natural studies and medicine. Chinese traders with scholarly interests were also actively engaged: they searched for classical texts long since lost in China but still available in Japan.

In the late eighteenth century, Japanese scholars interested in Qing classical studies—like their learned Korean counterparts—adapted the evidential research techniques pioneered by Qing literati. Sometimes this transmission occurred through Korea's frequent contact with the Qing court via tribute missions sent to Beijing. More often it occurred when Tokugawa scholars received via Nagasaki the most recent classical works published during the Qing dynasty.[36]

To some degree, what we see in this international exchange of books and knowledge in early modern China, Korea, and Japan is the emergence, before the arrival of the modern Western powers, of an East Asian community of classical scholars. In the late eighteenth and early nineteenth centuries, these scholars learned and adapted the evidential techniques pioneered in South China. Jesuit translations, particularly in the sciences, were part of this trade, but, except for Dutch Learning translations produced in Tokugawa Japan with the help of Dutch emissaries, physicians, naturalists, and traders housed on the Nagasaki harbor island of Deshima, Europeans did not transmit works from the Newtonian age until the middle of the nineteenth century.[37]

Imperial Support for Libraries and Publishing

In addition to the literary and scholarly projects that they sponsored, the Kangxi and Qianlong emperors actively promoted the palace collection housed in the Forbidden City, which was also the location of the Imperial Printing Office. Although an invaluable collection of books from the Ming imperial library had survived the Manchu conquest, many items in the Ming collection had been lost, when Beijing fell first to rebel and then Manchu forces. Losses included two out of the three complete sets of the major early Ming repository of ancient works from antiquity to the fifteenth century. Later, evidential scholars used this repository to reconstruct traditional Chinese mathematics.

More than fifteen thousand books in all genres of learning were published under the Kangxi emperor's patronage. Court scholars also compiled and published encyclopedias and dictionaries under imperial sponsorship during this time. Casting of a million and a half letters and symbols of movable type was required for one of them (Figure 3.2). Some Jesuits may have supervised the manufacture of the copper type, but their supervision may not have been necessary: the Koreans had expertise in movable-type printing.[38]

The Qianlong emperor's Imperial Library (SKQS) project, which was designed to preserve existing literature on all subjects and was later used to ferret out prohibited books, set off a flurry of book hunting by private individuals and official scholars in the 1770s and 1780s. Consequently, prices for rare books rose sharply. The first manuscript copy, which took up over 36,000 rolls (it was not printed for mass distribution because only seven copies were ordered), was finished in 1782 and housed in the Imperial Palace. Three additional sets were completed and housed in imperial sites. Essentially closed to literati and the public, these collections were maintained for imperial prestige. Later, additional copies were housed in the Yangzi delta. The Qianlong emperor and his editors justifiably thought that by preparing an accurate and permanent copy of the most important writings from antiquity to the present, including many Jesuit writings, they were making a significant contribution to Chinese literature and learning.

Between 1774 and 1794, many of the rare works discovered by the Imperial Library staff were printed using movable wooden type in the Imperial Printing Office. Subsequently, all the books printed in this manner were brought together to form a single collectanea consisting of 138 items. To show his appreciation to the Yangzi delta bibliophiles who had helped make the Imperial Library project a success, the Qianlong emperor ordered three more sets of the complete collection. These three libraries were opened to students and scholars possessing the required literati credentials.[39]

Commercial development in the Yangzi delta during the Ming and Qing dynasties was to eventually enable nineteenth-century printers to publish works on Western studies and modern science on an unprecedented scale. Moreover, printing helped transform the conditions under which texts in late imperial China were produced, distributed, and read. The print culture that emerged made possible the forms of private library collecting and book exchanges described earlier, and the scholar-printers provided the academic community—including its schools, academics, and libraries—with access to more specialized works than ever before. Such internal dissemination served as a precondition for the adaptation of Western science after the First Opium War.

Missing: Europe's Analytic Style of Mathematical Reasoning

Initially, the Jesuits played a significant role in the scientific developments associated with European academies of science. French Jesuits introduced the work of Isaac Newton (1642–1727) to the Chinese in their 1742 edition of the *Supplement to the Compendium of Observational and Computational Astronomy*, which mentioned Newton by name but gave no systematic presentation of his theories. By the late eighteenth century, however, the leading lights of Western Europe were ridiculing the Jesuits even before the pope abolished the order. The Frenchmen Joseph-Louis Lagrange (1736–1813), Jean Le Rond d'Alembert (1717–1783), Marquis de Condorcet, and Jean-Baptiste-Joseph Delambre (1749–1822) integrated mathematics and science and were among the first to apply the differential and integral calculus to mechanics and probability theory. The calculus of probability, for instance, provided a mathematical model for evaluating the validity of individual opinions and determining the probable outcomes of individual actions. The Jesuits did not transmit these new currents in Western Europe to the Chinese.[40]

The analytic style in European mathematics stressed a formal examination of the steps used in algebraic reasoning: that is, a problem was declared analytic when it could be solved algebraically via a theory of equations. This approach not only facilitated a firm foundation for the growth of statistics, but also led to the application of mathematics to late-eighteenth-century public policy in Europe. Mathematization of the social, economic, and political world in the name of political economy became de rigueur.[41] Condorcet, for example, who had been educated when French science confirmed and elaborated Newton's mathematical principles, sought to extend that mathematical vision into a universal social mathematics. These trends in social science, like the calculus, did not penetrate China in the eighteenth century.[42]

The analytic geometry of René Descartes (1596–1650) had translated and extended the classical art into a more flexible algebraic form. His equations did away with the special status of celestial circular motion, sublunar natural motion, and many other fixtures of the Aristotelian tradition to which the Jesuits still clung. Newton's "fluxions" provided a next step in the process. Although he did not use them in his 1687 *Principia,* which relied on formal geometric proofs, his fluxions represented a new notational form for a mathematics of flowing numbers (that is, numbers in "flux" over time). Newton's geometric proofs were thus passable for presentation, but useless for research. Continental European scholars rectified this problem by verifying and extending Newtonian natural philosophy using Leibniz's more effective calculus notations, which were based on an infinite series of small differences (that is, infinitesimals).[43]

In French hands, the sciences of algebra and calculus provided analytical tools that replaced static geometric presentation. For the mathematician Pierre-Simon Laplace (1749–1827), the task was to translate mechanics into a quantifiable algebraic language that explained movements in space over time. Mathematically inclined astronomers filled in the details of Newtonian gravity and the solar system. Adapting the calculus required differentiation and integration, and the summing of divergent series. French success created a sense of progress in explaining physical operations of the solar system, and the rule-like equations that Laplace, Lagrange, and others devised augmented the engineering toolkit for Newtonian mechanics.[44]

Newtonian science created laws of mechanics that superseded Renaissance mechanics, which had existed as a body of principles explaining the use of levers, wedges, and pulleys through rules of practice. European artisans of that period—like their Chinese counterparts—had understood the workings of machines but never considered developing an ordering theory or a set of principles to give their mechanics coherence. Classical mechanics had included the parallelogram of forces, the laws of the lever, principles of virtual work, and other diverse principles, but the Newtonian theory of force unified them into a new system in which mechanical devices were models for processes in nature. After Newton published his *Principia,* continental mathematicians regularized and systematized mechanics, pneumatics, hydrostatics, and hydrodynamics.[45]

Artisans quickly recognized that transformations in the mechanical philosophy would affect mechanical expertise, particularly in England. Engineers and mechanical artisans thus popularized the reform of natural philosophy in eighteenth-century England. By 1750, textbooks made the application of mechanical principles accessible to anyone literate in English and French, and artisans and engineers applied them. None were translated into Chinese, because the Jesuits never made the jump to the mathematicization of practical mechanics. They still accepted the Aristotelian notion of movement based on an object's own elemental makeup.[46]

By the late eighteenth century, experimental physics and mechanics had surpassed the use of geometry in Europe. Meanwhile, however, the Chinese were still using as a basis for mathematical learning the Ten Mathematical Classics (albeit a version that had been enhanced with Jesuit models for algebra). Such numerical methods may have facilitated mastery of the calculus by Chinese mathematicians, but this great leap forward was not to occur until the nineteenth century, about a century after Europe had made this crucial step.

Newtonian Science in England

As Europeans extended Newton's *Principia*, entrepreneurs and engineers employed mechanics to link natural philosophy with the practical arts, science, and technology. The emerging "engineer's toolkit" mathematized the laws of nature by elaborating the calculus. The new science was integrated differently in the British social and cultural landscape than on the Continent. From the 1720s, British public schools taught basic mathematics, including algebra, geometry, surveying, mechanics, and astronomy. Arithmetical and mathematical texts doubled in number during the first half of the eighteenth century. Young artisans in the 1720s already knew rudimentary mathematics and mechanics. At the same time, Oxford and Cambridge in their somnolence included only classical geometry in the curriculum.[47]

During this period, science education was widespread and often informal: the British learned of new scientific developments through academic discussions, lectures, textbooks, and newspapers in a process that was to help spark the Industrial Revolution in Britain. By 1750 England had many male entrepreneurs who approached the production process mechanically. Production could be mastered using machines and conceptualized in terms of weight, motion, and the principles of force and inertia, which workers in factories applied. Aspiring inventors knew that the practical applications of mechanics allowed engineers using the calculus to quantify the motion of fluids and solids, measure the weight and pressure of different substances, and create mechanical devices such as pumps, pulleys, levers, weights, as well as start to learn how to utilize heat, electricity, and light.[48]

Newtonian physics, in contrast with the doctrinal inflexibility of the Catholic Church, Jesuit colleges, and fundamentalist Calvinists, quickly developed into the scientific core of natural religion. By the 1720s, Newtonianism had become the body of natural learning for laymen in Britain to master, practice, and apply. One of them, James Watt (1736–1819), modified and improved the simpler steam engines of the eighteenth century and turned them into the most advanced technology of the age. The steam engine linked Newtonian mechanics and engineers in a cycle of manufacturing and industrial advances that soon led to the efficient use of natural deposits of fossil fuels and thus an unprecedented amount of new energy for practical work in factories, arsenals, and on ships.[49]

Watt's engine became a model for industrial change, and his own intellectual life reflected the cultural origins of the first Industrial Revolution. Watt came from a family of mathematical practitioners who were well informed about instruments and machines. He understood the latest mechanical science and turned to mechanized industry as an engineer and entrepreneur. A new industrial mentality emerged in the eighteenth-century Britain, which was marked by a set of recognizable values, experiences, and learning patterns among engineers and industrialists. The autonomy of the press in England and its independent public favored the interests of practical-minded scientists and merchants with industrial interests. After 1750, laymen and civil engineers communicated through a common scientific heritage that fashioned the mental world of the Industrial Revolution.[50]

The French Century in Science and Engineering

By the 1790s, the French were emulating the British educational system, although as of midcentury they had already become leaders in those applied mechanics requiring mathematical training and basic geometry. The ministry of A. R. J. Turgot (1727–1781) drew from both science and systematic knowledge to revitalize the French monarchy in 1774. Indeed, the Paris Academy of Science in the eighteenth century was more prestigious in scientific circles than was the London Royal Society. Science in France evolved further toward the formation of a profession whose expertise presupposed mastery of a technical body of knowledge that granted social prestige. As the profession of science gained its own jurisdiction over the education, qualifications, and conduct of its members, its disciplines approached the levels already reached by the divinity, law, and medical fields in French universities.[51]

As loyal Aristotelians, Jesuits fought Newtonianism in their colleges into the 1740s and 1750s. After that, however, the curriculum of nearly four hundred French colleges shifted from Cartesian materialism based on corpuscles to theoretical and applied Newtonianism. If Cartesian materialism was used by some clerical supporters to glorify Louis XIV, the negative associations of its links to absolutism and atheism were overcome in France when Voltaire (1694–1778) and the philosophes offered Newtonian science and English society as models of enlightenment. They also attacked Cartesian science as insufficiently experimental.

Unlike in Britain, major industrialization did not occur in France until the nineteenth century, despite Newton's influence on the emerging French scientific community. Abbé Jean Antoine Nollet (1700–1770) was one of the chief French promoters of the new science and its mechanical applications. He learned the techniques of demonstration from the Dutch Newtonians and then established a series of physics lectures while traveling in France. Such French popularizers of mechanics provided an alternative to the Cartesian physics that had been dominant since the 1690s at colleges such as the University of Paris. This physics also filtered into the 1750s volumes of the *Encyclopédie* by Denis Diderot. Although the inspiration was Baconian, the volumes were filled with drawings and descriptions of mechanical devices. Some twenty-five thousand copies circulated before 1789, but none in China.

While England evolved toward a more balanced polity, the French philosophers justified the absolutist state and its support for scientific inquiry (a perspective that appealed to Qing emperors such as Kangxi). Condorcet, for example, thought that science academies benefited the monarchy. And although Diderot and d'Alembert, among others, urged that mechanics be the first science studied for its rather egalitarian utility, this vision was realized only after the French Revolution had overturned the aristocratic domination of science academies and military engineering schools.[52]

In the 1750s, the French monarchy focused on steam-powered boats for military use. In the 1770s and 1780s, it encouraged mechanical devices for controlled farming. In the French system, engineers and men of science served the state (in Britain, by contrast, science was geared more toward private entrepreneurs). An emphasis on mechanics in French education began in French colleges in the 1770s and 1780s, but French military engineers were the only cluster with sufficient mechanical knowledge. Their values prioritized military needs over commerce. Laplace, for example, arrived in Paris in 1769 to take up his first teaching post at the Ecole royale militaire.[53]

Civil and military engineering were introduced as fields of study in schools for the state's corps of civil engineers, which was given royal patronage in 1775. In the 1780s, the corps consisted of 230 commissioned engineers whose work focused on designing, drafting, and verifying structures. The Ecole des ponts et chaussées became a professional school with a technical curriculum, which was introduced via French engineers to China at the Fuzhou Naval Yard in the 1860s and 1870s. Subjects, arranged in descending order of importance and difficulty, included:

1. *Mathematics:* mechanics, hydraulics, and differential and integral calculus; algebra and conic sections; and elements of geometry.
2. *Architecture:* bridges; docks, jetties, locks, dykes, and canals; civil building; and stonecutting (stereotomy).
3. *Style and method:* drafting, the application of leveling and volumetric calculations to earthworks, and the estimation of work tasks.
4. *Drawing:* maps (geographical and topographical), figures, and landscapes.
5. *Writing:* block lettering and penmanship.[54]

Meanwhile, in Qing China, the Yellow River Conservancy, which for much of the eighteenth century included hundreds of officials in several provinces and thousands of military technicians, continued to rely on traditional hydraulic technologies for river and flood control along the North China plain. By the middle of the nineteenth century, the Qing state had disbanded this conservancy out of frustration that recent flooding along the Yellow River was too costly and too unmanageable.[55]

In 1793, at the height of the French Revolution, all scientific academies were abolished in France. Not until 1795 was the Paris Academy revitalized, and then with largely new personnel because it had lost half of its members. New populist versions of science filled the void left by the disestablishment of French Enlightenment science. Laplace, for example, who was appointed in 1795 to give mathematical lectures at the Ecole normale, ridiculed Leibniz's proof for the existence of God.

The triumph of engineering as the quintessential practical science was evident in the fanfare surrounding the founding of the Ecole polytechnique in 1794. It became an incarnation of revolutionary ideals and its new vision of science to change the world. The leaders of the Revolution and Napoleon all embraced an industrial vision for France that drew on the power of science to change society and nature. French civil engineers were prized a generation after their English counterparts in the 1760s and 1770s. In the nineteenth century the Napoleonic success in science and technology was studied and transferred to Germany.[56]

Meanwhile in China, when members of the Macartney mission visited the imperial glass-making workshop in 1793–1794, one of them noted that the site was neglected. After Gabriel-Leonard de Brossard's (1703–1758) death in 1758, production at the Jesuit workshop had declined. Subsequently in 1827 the Qing court confiscated all missionary property, and in 1860 British troops, who had taken Beijing and forced the imperial court to flee to their summer retreat, destroyed the emperor's Lofty Pavilion Garden-Palace.

Macartney's German-made planetarium attracted some interest, but his uninformed view of the Chinese convinced him that diplomatic success would naturally follow once British cultural superiority and scientific expertise had been demonstrated to the Manchu emperor. (I should add that Macartney, although the British East India Company sent him, was aristocratic to the core, and seems not to have sensed how the English Industrial Revolution was relevant for China.)[57]

The Macartney Paradox

When they arrived in China in 1793, neither Macartney nor his ship's mechanic and mathematician cum astronomer James Dinwiddie (1746–1815) thought much about the fact that the British East India Company had purchased the German-made planetarium at an auction and coated it in oriental-style ornamentation for the Chinese market. Only when he visited the lavish Qing imperial gardens filled "with spheres, orreries, clocks, and musical automatons of such exquisite workmanship" did Macartney stop to consider the limits of his scientific apparatus. Afterward, when someone suggested that the Chinese might be more interested in British machinery than the German-made planetarium, the comment was set aside out for fear that the clever Chinese would quickly learn how to copy and export such machinery.[58]

When the British presented the Qianlong emperor with a model of the warship *Royal Sovereign,* for example, he asked some mechanical questions that reveal his close interest in the artillery. Indeed, Macartney never presented the pulleys, air pump, chemical and electrical contrivances, or even the steam-engine models that he had on board. Nor did the mission present the chronometer that Macartney had brought as a possible gift. A new means to determine longitude, the chronometer would have been more efficient than the Jesuit method for surveying that the Manchus were using to appraise their domains.

Instead the apparatuses were returned to the British East India Company or given to Dinwiddie, who lectured on them and presented some experiments in Guangzhou to the English factory, which was attended by Chinese merchants. Macartney remarkably noted: "Had Dinwiddie remained at Canton and continued his courses, I dare say he might have soon realized a very considerable sum of money from his Chinese pupils alone." Chinese merchant interest in Dinwiddie's experiments and contrivances in 1793 anticipate the period after the Opium War, when many Chinese quickly took notice of advances in engineering brought about during the Industrial Revolution.[59]

Given the sophistication of Chinese manufactures in books, paper, porcelain, silk, and cotton handicrafts for some 350 million Chinese by 1800, why have historians of science favored a European-dominated story of the rise of modern techno-science? Part of the answer lies in the claims made by the European latecomers to the Asian trading markets about their nineteenth-century successes—accomplishments that featured dramatic advances in early modern capitalism and Newtonian science. Modern economic growth, that is, the revolutionary growth in productivity per capita in Western Europe versus in China, may be another part of the answer. But surely those who were first to embark on high-volume industrial production and international trade—India and China—have a different perspective on this irony derived from the law of diminishing returns and the lag in adopting Newtonian science.[60]

SCIENCE AND THE PROTESTANT MISSION

After the French Revolution and Napoleonic Wars, Christian missionaries again took the lead in Sino-European interactions. The victories over Napoleon at sea in Egypt and on land at Waterloo validated Great Britain's global importance. By 1820, the British Empire controlled a quarter of the world's people, many of whom were colonized between the 1756 Seven Years' War and 1815. Moreover, the American Revolution had redirected Britain's expansionist aspirations toward south and southeast Asia and China.[1]

British Protestants—secular and religious—perceived the Qing empire as an obstacle to open commerce and Christian evangelism. English politicians, merchants, and evangelicals articulated through religion and science their goal of opening China through diplomacy to the international community and enlightening its peoples. But when Manchus and Chinese resisted the goals of the Macartney mission in 1793 and the Amherst mission in 1816, the British grew anxious over whether the diplomatic approach would give them access to the most important market in the world. Moreover, between 1828 and 1836 more than $38 million flowed out of Qing China to pay for illegal opium imports sponsored principally by the British East India Company—a situation that outraged free-trade supporters and evangelists. Their call for free trade, fueled in part by the influential book *Wealth of Nations* (1776) by Adam Smith (1723–1790), climaxed in England in 1833 when, after four years of politicking, free traders and their evangelical supporters convinced Parliament to abolish the East India Company monopoly in China.[2]

Even after the British victory in the First Opium War (1839–1842), which led to highly favorable trade agreements for Britain and a resumption of the flow of opium to China, Western missionaries had only limited influence in China. Yet Protestant missions were significant for the introduction of modern science: they principally funded the new translations, newspapers, and schools that introduced it in the 1850s. This effort was interrupted during the massive Taiping conflagration, when from 1850 to 1864 anti-Manchu and anti-gentry discontents rose up in a highly destructive peasant rebellion that significantly changed the tenor of scientific education and inquiry once it had been quelled using new Western armaments. As we will see in Chapter 5, from the 1860s on the impetus for science and technology shifted from the Protestant missions to the reforming Qing state and its new Western-oriented policies and institutions.

Protestant Missionaries in China

Once the East India Company decided to permit missionaries into its territories, Protestant denominations quickly organized. Similar groups organized in mainland Europe, and even the Jesuit order regrouped to enter China again. Officially nondenominational, the London Missionary Society (LMS), founded in 1795, played an important role in China throughout the nineteenth century. Similarly, the Church Missionary Society was organized in 1799 as an independent voluntary organization under the Church of England. The "Evangelical Revival" in Britain and the "Great Awakening" in the

United States spawned new denominations such as the Methodists and new organizations such as the Salvation Army, the Sunday school, as well as the YMCA and YWCA.[3]

Other Protestant missionary societies organized throughout Europe, although the English presence was felt most in China. Similarly, the American Board of Commissioners for Foreign Missions organized in 1810. Extraordinary wealth and energy generated in the aftermath of the Industrial Revolution in England, the United States, and in continental Europe was channeled overseas toward the missionary movement. Although Britain and the United States remained political antagonists, the religious controversies of the nineteenth century were common to both. The religious revival that swept through Great Britain in this period also swept through the churches and universities of New England.[4]

The Protestant mission in China laid the foundation for future work by arranging a corpus of Christian literature for printing. It also provided groundwork for gathering books in Chinese by the missionaries. Robert Morrison (1782–1834), for example, who worked for the LMS, put together an impressive library collection that included works on traditional Chinese medicine and mathematical astronomy. Morrison's collection contributed to the development of LMS Press. Established at Malacca in 1818 at the Anglo-Chinese College, the LMS Press later moved to Hong Kong and then to Shanghai in 1842, where it was renamed Inkstone Press. After 1847, when Alexander Wylie (1815–1887) began to supervise its operation, Inkstone Press became the publishing mecca of missionary activities.[5]

In the 1830s, missionaries and their Protestant presses reached South China. The first leading missionary publication in China was the *Chinese Repository*, a monthly begun in Guangzhou in 1832 by the American Elijah Coleman Bridgman (1801–1861). He was joined in 1833 by Samuel Williams, and the journal became the main outlet for serious Western scholarship in China until the 1850s.[6]

Among medical missionaries, the American Peter Parker (1804–1888) organized the Medical Missionary Society in Guangzhou in 1838. The German Karl Friedrich August Gützlaff (1803–1851) and Bridgman were active in Singapore and Guangzhou and served as secretaries when the joint missionary-merchant publication *Society for the Diffusion of Useful Knowledge in China* was founded in November 1834. In 1837, the Diffusion Society initiated an ambitious plan to present in Chinese works on history, geography, natural history, medicine, mechanics, natural theology, and other subjects. For this series, Bridgman produced a treatise on the United States that was printed in 1838, revised in 1846, and expanded for publication in 1862. It influenced Qing officials eager to understand the new maritime world they faced after the First Opium War.[7]

Gützlaff published the *Eastern-Western Monthly Magazine* in the 1830s in Singapore and Guangzhou, which the staff of Lin Zexu (1785–1850) monitored and translated when Lin arrived in Guangzhou in March 1839 as the imperial plenipotentiary to solve the opium crisis in South China. Lin's *Gazetteer of the Four Continents* included portions of Gützlaff's magazine, which was one of the major sources for Wei Yuan's (1794–1856) influential *Treatise on the Maritime Countries*—a work that was initiated in 1841, published in 1844, enlarged in 1847, and doubled in size in 1852.[8]

After the First Opium War, the treaty between Britain and the Qing dynasty opened the South China coast to Euro-American trade—and the ports of Guangzhou, Shanghai, Fuzhou, Amoy, and Ningbo to foreign residence. The island of Hong Kong became a British colony, and Christian missions from several European states in Dutch Batavia, English Malacca and Singapore, and Portuguese Macao quickly moved their missionaries and printing equipment to the South China coast. After the Sino-French Agreements of 1858 and 1860, France assumed the role of protector of Roman Catholic missions in China.[9]

Modern Science and Medicine in South China

Wei Yuan's (1794–1856) famous *Treatise on the Maritime Countries* contained a plan for Qing military defense that included construction of a navy yard and arsenal near Guangzhou where Westerners could teach the Chinese how to build ships and manufacture arms. The Manchu governor-general in Guangzhou, Qigong (1777–1844), called for including mathematics and manufacturing in Qing civil examinations in the early 1840s. Because such requests fell on deaf ears in Beijing, the early introduction of modern science and technology was left to the Protestants and their converts in South China.

Daniel Jerome Macgowan (1814–1893) and Benjamin Hobson (1816–1873), both physicians, were the key pioneers in the late 1840s and early 1850s. The American Macgowan initially served as a medical missionary in Ningbo. Later he became a freelance lecturer and writer, as well as a member of the Qing maritime customs service. After moving to Hong Kong, Hobson, an English medical missionary, pioneered a series of medical and science translations, coauthored with the Chinese, for his premedical classes in Guangzhou. Meanwhile, Alexander Wylie came to Shanghai from England to join the LMS Inkstone Press as its printer. He had been selected by James Legge (1815–1897), who pioneered the translation of the Chinese Classics at Oxford with the help of Wang Tao (1828–1897).

While in Ningbo, Macgowan published the *Philosophical Almanac* (1851), a scientific work devoted mainly to electricity and electrotherapy. In it, Macgowen used the electrolysis of water to demonstrate that it was a composite substance and not a single, unified agent. He also introduced what many thought was the curative power of electricity via electrodes attached to the body. Both he and Hobson regarded electricity as a newly discovered key to understanding the nervous system.[10]

Similarly, Hobson prepared the *Treatise of Natural Philosophy*

(1851), associating science with the Chinese tradition of "broad learning about things" *(bowu)*. Initially published by the Guangzhou Hospital, Hobson's translation drew on Macgowan's Ningbo *Almanac* but became the more influential when reissued in Shanghai. Wylie and Li Shanlan preferred calling science "the investigation of things and extension of knowledge" in their scientific translations for the Inkstone Press. As with Jesuit translations of Aristotelian natural studies, Chinese co-workers were still influencing the choice of terms in Protestant scientific translations. In the 1850s, these Chinese terms for the sciences were quickly transmitted to Japan.[11]

Western Anatomy and Traditional Chinese Medicine

Hobson also produced a series of other works to educate his medical students, including the *Summary of Astronomy* (1849), *Treatise on Physiology* (1851), *First Lines of the Practice of Surgery in the West* (1857), *Treatise on Midwifery and Diseases of Children* (1858), and *Practice of Medicine and Materia Medica* (1858). Hobson's work the *Treatise on Physiology* presented modern anatomy (Figure 4.1) and reintroduced the centrality of the brain and the nervous system, which Jesuits had tried unsuccessfully to do in the late Ming and early Qing.[12]

This unprecedented series of modern medical works remained standard in China until the late nineteenth century, and each volume was reproduced widely in Japan. Later John G. Kerr (1824–1891), John Dudgeon (1837–1901), and John Fryer (1839–1928) translated new texts on medicine that superseded Hobson's works. The Medical Missionary Association was formed in 1886 and printed its own medical journal in Chinese.[13]

The missionaries believed that medicine was at a low ebb in China. Yet when Hobson translated Western medical works into classical Chinese, the heat-factor tradition for dealing with fever-

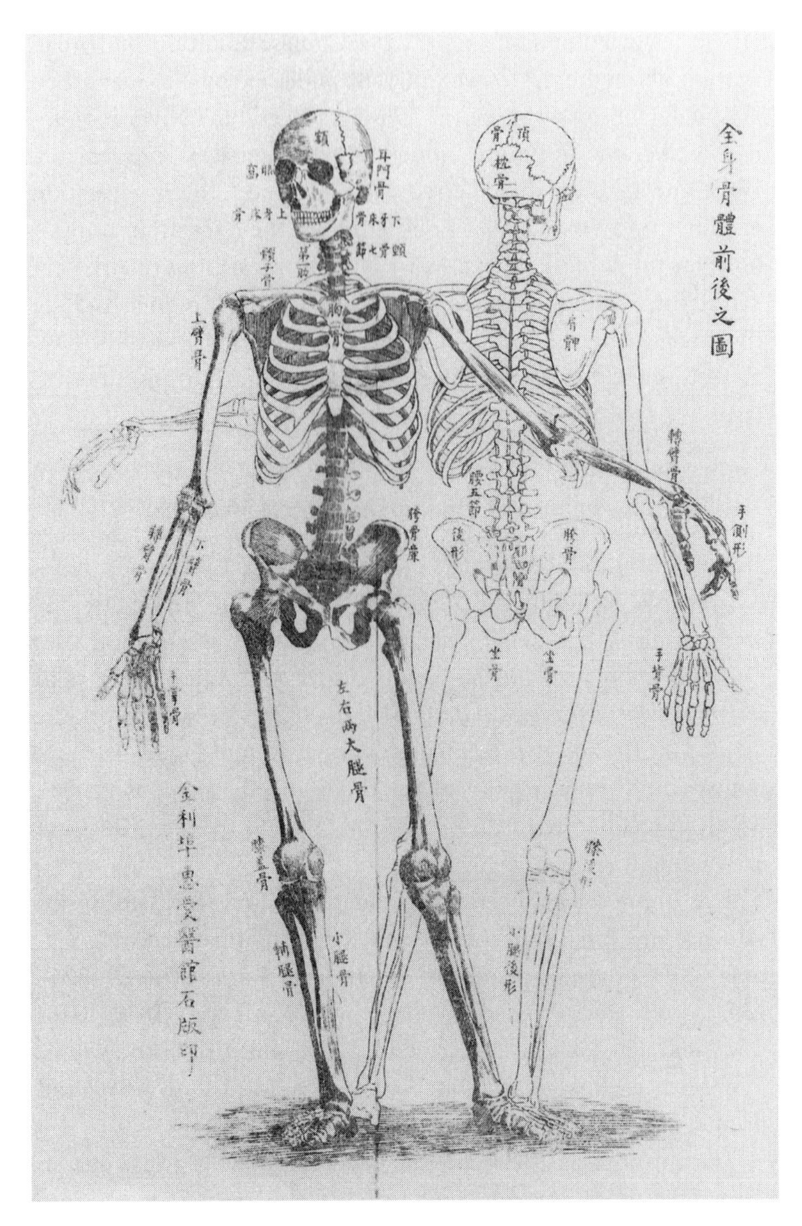

Figure 4.1. Hobson's "Frontal and Rear View of the Human Skeleton," in the *Treatise on Physiology (Quanti xinlun). Source:* Benjamin Hobson, *Quanti xinlun* (Guangzhou: Hum-le-fow Hospital, 1851).

inducing illnesses that had emerged in China in the seventeenth century was growing increasingly prominent among traditional physicians in South China, where the missionaries were often assigned. Regional traditions dealing with southern infectious diseases and northern cold-damage disorders continued to evolve in the nineteenth century. In the process, heat-factor illnesses became a new Chinese category.[14]

The mid-nineteenth-century emergence of a medical tradition stressing heat-factor therapies coincided with the introduction of Western medicine in the treaty ports, particularly Guangzhou, Ningbo, and Shanghai. The newly systematized canon for heat-factor disorders was constructed from many scattered writings on heat-factor illnesses in the ancient medical literature and from more recent works produced mainly by early Qing physicians. Wang Shixiong's (1808–1890) *Warp and Weft of Warm and Hot Factor Diseases,* for example, was published in 1852, a year after Hobson's *Treatise on Physiology* appeared.[15]

In *Warp and Weft,* Wang acknowledged the usefulness of Hobson's anatomical depictions of female reproductive organs and human physiology to improve diagnosis, but he subordinated such new information to the traditional therapeutic regime.[16] The Chinese had long accepted new understandings of human anatomy when they could assimilate them within the Chinese focus on internal conduits of qi: an ancient dynastic history, for instance, recorded a complete human dissection. Moreover, Song physicians had mapped acupuncture and moxibustion therapy onto the skeletal body, and the internal organs had also been drawn and modeled. Illustrations of internal organs drawn from a cadaver after an execution were widely available in 1113. Forensic medical officers later applied this knowledge of internal organs to crime investigations.[17]

In this intellectual environment, both traditional natural studies and Chinese medicine faced challenges. Chinese medical efforts to treat southern infectious illnesses paralleled the gradual emergence of tropical medicine during the late nineteenth century, when the British Empire was increasingly populating the tropics with its own physicians. These networks of doctors and their medical reporting system from Africa to India and South China in turn addressed interregional infectious diseases such as malaria. Colonial physicians cumulatively sent back information about epidemics and infectious illnesses to London, the metropole of global medicine.[18]

Patrick Manson (1844–1922), a port surgeon and medical officer in the Imperial Chinese Customs Office since 1866, helped establish the London School of Tropical Medicine in 1898. Assigned for over two decades to Chinese treaty ports, Manson studied tinea, Calabar swelling, and blackwater fever before he developed a focus on tropical hygiene. He distinguished himself with his research on filariasis, a disease endemic in South China for which neither Chinese nor European medicine had a remedy. In particular, he observed in 1878 that the filariae worms causing elephantiasis passed part of their natural life cycle in the *Culex* mosquito, thus demonstrating transmission by parasites and explaining their natural history.[19]

Until the idea was unseated by the germ-parasite theory of disease in the late 1890s, Europeans regarded malaria as a miasma defined by human fever; Hobson himself associated malaria with putrid air. In the latter half of the nineteenth century, Western physicians tried to explain such extreme fevers by using a chill theory that described tropical illnesses according to the degree of change in an individual's physiology. Hot days and cold nights produced such fevers, most thought. Such views overlapped with Chinese notions of cold- and heat-factor illnesses, within which southern miasmas were identified medically as the causes for infectious diseases spread by the wind. During this time, the Chinese were also investigating specific illnesses such as leprosy and malaria associated with South China.[20]

This demarcation of illnesses unique to North or South China meant that cold-damage therapies were inappropriate for southern illnesses. Moreover, these medical currents represented a critical test for a therapeutic tradition cobbled together from the medical classics. The challenge of Western medicine in the nineteenth century, and its increasing focus on parasitic causes of tropical diseases, added to the native debates of late Qing medical practitioners.[21]

The Chinese increasingly acknowledged the need to synthesize Chinese and Western medicine. Ding Fubao, for example, linked cold-damage disorders to the specific illness that Western physicians identified as typhoid fever. Germ theory was added to discussion of warm- versus cold-factor illnesses. And Chinese physicians began to explain the wasting of the body's natural vitality in terms of tuberculosis (in the sense of a wasting disease) and gonorrhea (as a depletion illness).[22]

Unlike Ming-Qing astronomy, which was completely reworked in the seventeenth and eighteenth centuries by the introduction of Western techniques, traditional Chinese medicine did not face a serious challenge from Europe until the middle of the nineteenth century. Except for smallpox inoculations, quinine therapy for malaria, and a number of herbal medicines unknown in China, the European medicine brought by Jesuit or Protestant missionary physicians did not achieve superior therapeutic results—that is, until a relatively safe procedure for surgery combining anesthesia and asepsis was developed at the turn of the twentieth century.[23]

But the translations Hobson prepared led some literati to question the use of traditional Chinese medicine in the nineteenth century. Xu Shou (1818–1884), one of John Fryer's collaborators, was one of the first scholars to attack the application of classical theory to medicine—particularly medical prescriptions stressing moral cultivation to calm bodily excesses.[24] In an article published in April 1876, Xu also attacked the yin-yang dichotomy and the five phases correspondence system. Moreover, Xu Shou complained that while the literati had integrated Western and Chinese mathematics, they had paid little attention to the strengths of Western medicine. Xu called for a similar synthesis of Western experimental procedures, linking chemistry and Chinese strengths in materia medica.

Meanwhile, outside the missionary hospitals and clinics in the treaty ports, Hobson's translations were not popular due to the Chinese distaste for surgery. Minor surgical procedures such as cutting warts, lancing boils, cauterizing wounds, removing cataracts, and castration for eunuchs were relegated to the nonliterati majority of physicians. Hobson's *Treatise on Physiology* and his *Treatise on Midwifery* introduced invasive surgery for childbirth, drawn from the anatomical sciences that had evolved in Europe since the sixteenth century. But although anatomy could pinpoint childbirth dysfunctions as happening in the uterus, interventions were dangerous even by Western standards until modern surgery integrated sterilization techniques with anesthetization procedures.[25]

Rather than invasive surgery for childbirth problems, Chinese physicians preferred practical therapies for women based on their holistic, interactive model of the human body. Wang Tao, then Hobson's associate, noted that Western treatises for women's medicine did not prescribe therapy for dealing with illnesses, which was the strength of procedures in the Chinese toolkit. To reply to this complaint, Hobson prepared a more practical guide in 1857 in which he presented the medical prescriptions left out in his earlier *Treatise on Physiology*.[26]

On the basis of his own examination of human corpses, Wang Qingren (1768–1831), one of the few Chinese physicians to take anatomy seriously, contended that all of the bodily depictions in the Chinese medical classics were inaccurate. His *Corrections of Errors in the Forest of Medicine* (1830, 1853) also maintained that the brain was the central organ of the body, a view that became more prominent after Protestant medical texts such as Hobson's were translated into Chinese.[27]

From Western Medicine to Modern Science

Hobson's work represented the first sustained introduction of the modern European sciences and medicine in the first half of the nineteenth century. His 1849 digest of modern astronomy, for instance, presented the Copernican solar system in terms of Newtonian gravitation and pointed to God as the author of the works of creation. Thereafter, Newtonian celestial mechanics, which was based on gravitational pull, was increasingly presented in Protestant accounts of modern science—with God's role central.

A natural theology also informed Hobson's *Treatise of Natural Philosophy*, which was the first work to introduce modern Western chemistry. The textbook presented the fifty-six elements, but Hobson presented God as the ultimate creator of all the myriad changes in things. Although later changed, Hobson's chemical terminology presented the names of gases in Chinese and implicated them in the chemical makeup of the world, thereby supplanting the four-elements theory of the Jesuits and challenging the Chinese notion of the five phases.[28]

By including sections on physics, chemistry, astronomy, geography, and zoology for his Chinese medical students, Hobson unexpectedly attracted the interest of literati unsuccessful in the civil examinations.[29] Fryer described a group of Chinese literati investigators who earlier had met to go over Jesuit works on mathematics and astronomy. They used Hobson's *Treatise* to catch up with findings since the Jesuits. This group, which included Xu Shou and Hua Hengfang (1833–1902), also carried out experiments. After fleeing the Taiping rebels in the early 1860s, they were invited by Zeng Guofan (1811–1872), leader of the victorious Qing armies, to work in the newly established Anqing Arsenal. Hua began translation projects with Alexander Wylie and Joseph Edkins (1823–1905), while Xu worked on constructing a steamboat based on Hobson's diagrams.[30]

Protestants and Modern Science in Shanghai

British and American Protestants in China recognized that Chinese literati were interested in the sciences that the missionaries accepted as part of their Christian heritage. Like their Jesuit predecessors, British and American missionaries viewed science as emblematic of their superior knowledge systems. Consequently, their introduction of modern science and medicine to China was not only a missionary tactic; it was a way of showcasing the wealth and power of Western nations.

Among treaty ports, Shanghai by 1860 was the main center of foreign trade, international business, and missionary activity. The LMS Press in Shanghai became the most influential publisher of Western learning after 1850. It published translations from a distinguished missionary community, which included Alexander Wylie and Joseph Edkins, and worked with outstanding Chinese scholars such as Li Shanlan and Wang Tao, who had moved to Shanghai after failing to gain a place in the imperial civil examinations. Wang went to Shanghai in 1849 and was invited to serve as the Chinese editor at the Inkstone Press.[31]

Alexander Wylie and the Shanghae Serial

In the 1850s, other Protestant journals published in Chinese, such as the *Shanghae Serial (LHCT)* at Inkstone Press, introduced new fields in the Western sciences. Beginning with the *Shanghae Serial*, the literati notion of investigating things moved from encompassing classical learning and natural studies to designating a specific domain of knowledge within the natural sciences. The Jesuits had not challenged the diffuse methodological focus of the native investigation of things. They simply added the fields of *scientia* to it. Through the Protestant translation work of Wylie, Li Shanlan, and others for the *Shanghae Serial*, however, the investigation of things increasingly demarcated the new Western natural sciences. For example, a scientist was now called "someone who investigated things and extended knowledge."[32]

Unlike the weekly *North-China Herald and Supreme Court and Consular Gazette*, which was founded in 1850 in Shanghai and eventually became the leading foreign language newspaper in China, the *Shanghae Serial* presented Western affairs and culture, new scientific fields, and Christian theology in classical Chinese. At Inkstone Press, Wang Tao and Li Shanlan, who had joined in 1852, were the chief Chinese translators. Inkstone soon was a meeting place for the many Chinese literati living in or passing through Shanghai who were interested in Western learning, such as the future diplomat Guo Songtao (1818–1891). Many like him would go on to distinguished careers serving the Qing dynasty after the imperial government embarked on a program of military, educational, and scientific reform in the aftermath of the Taiping Rebellion.[33]

A talented missionary printer and translator, Alexander Wylie produced the *Shanghae Serial* monthly for thirteen issues in 1857 (there was an intercalary month in the Chinese lunisolar calendar), and for two more issues in 1858, before it suddenly stopped on June 11. Wylie was also busy during this period making some remarkable inquiries about Chinese science and mathematics with the help of Li Shanlan. Through this interaction, Li successfully completed the transition from the traditional craft of algebra to understanding the modern calculus.

Around 1860, Wylie and Li also started but left unfinished a complete translation of Newton's *Principia*. It presented for the first time in Chinese his laws of motion. About 1868, Li Shanlan completed the translation in the Jiangnan Arsenal's Translation Department with the help of John Fryer, but the work was never published because it was too hard to understand. After 1899, the manuscript was lost. The title for the Li-Fryer version of the *Principia* in Chinese meant "Investigating and Extending Knowledge of Mathematical Principles."

Wylie's and Li's 1859 translation of *The Outline of Astronomy* (1851) by John Herschel (1792–1871) grew out of their early collaboration. In his astronomy, the Cambridge-educated Herschel moved away from the work of late-eighteenth-century Newtonians who had stressed geometrical demonstrations over algebraic processes. In their writings, Herschel and his generation of British scholars contradicted the static model of the solar system (demonstrated by, for example, the planetarium that Lord Macartney had thought a grand gift for the Qianlong emperor in 1793), and moved toward a more powerful, dynamic mathematical physics by applying partial differential equations to celestial motions. In effect, Wylie and Li's translation of *The Outline of Astronomy* presented a British version of the French Newtonians.[34]

In 1855, Wylie and Li Shanlan also translated books seven through fifteen of Euclid's *Elements of Geometry*, which had been left out of the Jesuit versions. When compared to Newtonian mechanics and the calculus, however, it had less relevance for industrial enterprises. In 1853, Wylie prepared his *Compendium of Arithmetic* with the help of his Chinese student. This primer contained the rudimentary rules of arithmetic, the theory of proportion, and logarithms, but Wylie also solved traditional Chinese mathematical problems to make the translation accessible.

Wylie used the *Compendium of Arithmetic* to refute the eighteenth-century Chinese claim that Jesuit algebra was the same as earlier single-unknown methods. Wylie and Li Shanlan also published through Inkstone Press an 1859 translation of the 1836 work *Elements of Algebra*, which in turn had been taken from the original by Augustus De Morgan (1806–1871), an 1827 graduate of Trinity College in Cambridge. Part of the first generation of mathematical reformers at Cambridge in the 1820s, De Morgan applied the calculus to mechanics, physical astronomy, and other branches of natural philosophy. British physicists followed De Morgan and sought mechanical models to quantify phenomena.[35]

Similarly, Wylie and Li stressed modern algebra as a mathematical language for the natural sciences. In particular, they related modern algebra to traditional Chinese mathematics by substituting it for procedures solving equations with a single unknown or four unknowns. Wylie emphasized that the Chinese "quadrilateral algebra" (that is, procedures involving four unknowns) was superior to the Jesuits' elementary algebra, and acknowledged that Western scholars had not studied the two traditional methods. Nevertheless, Li and Wylie also refuted the theory that the science of algebra had originated in China.[36]

Religious and Scientific Discussions in the Shanghae Serial

Five thousand copies of the first issue of the *Shanghae Serial* were printed in January 1857. Most of the subscriptions were ordered by Shanghai foreign residents, while subscriptions from Shanghai Chinese made up only 8 percent. The journal was also available in five treaty ports and Hong Kong. Another 11 percent of the subscribers were Chinese and foreigners from Fuzhou in the southeast. The print run for issues declined gradually from a peak of 5,190 copies in March 1857 to 2,500 copies for the last two issues in 1858. It was widely read in Japan, where prefectural libraries ordered it along with other translations from China.

As the journal's print runs diminished, the editors added a series on mechanics and then on popular mechanics to interest a less exclusively literati audience. The rest dealt with a balance of theological and religious issues, the core message of the LMS. The scientific coverage represented Wylie's influence, and political-commercial events drew readers from the business community. The first issue opened with a "Brief Introduction" in which Wylie stressed the value of science for understanding the world, and introduced the Chinese terminology for its most important fields: chemistry, geography, animals and plants, astronomy, calculus, electricity, mechanics, fluid mechanics, and optics and sound. Wylie also mentioned Chinese traditional approaches to single-unknown mathematics, then discussed the importance of algebra and the calculus for astronomical calculations.[37]

Because the *Shanghae Serial* was an LMS publication, Wylie quickly turned in his introduction to natural theology to elaborate the providential design of the "maker" in the universe: "All things that are born and mature on the earth are made by the Lord on High, and when we examine the reasons, these also result from the Lord on High." In this way, the *Shanghae Serial* celebrated God, creation, and salvation. Furthermore, God had created sixty-four (not fifty-six) chemical elements, which were not a creation of natural forces. We will see similar adjustments to Darwinian evolution.[38]

Protestant theology in Shanghai bore the imprint of William Paley (1743–1805), whose combination of theology and utilitarianism also influenced Charles Darwin (1809–1882), Paley's student at Cambridge. Paley's work stressed the "Evidences of the Existence and Attributes of the Deity, Collected from the Appearances of Nature," which was a major theme that informed most scientific texts translated by missionaries in China after 1850. Until 1895, Protestants and Catholics ignored Darwin's theory of biological evolution based on natural selection.[39]

Beginning in the *Shanghae Serial*'s fifth issue of 1857, and continuing until June 1868, Wylie and Wang Tao published their Chinese translation of the article "Progress of Astronomical Discovery in the West." It opened with a theological refrain concerning God's powers but quickly addressed ancient astronomical traditions in China, Egypt, India, and Mesopotamia. According to Wylie, Chinese observances of solar eclipses recorded during the Shang dynasty were the most ancient. In Chinese, Wylie and Wang presented the man of science in light of the ancient Chinese category of a computist, which drew on the equation of computists with mathematical astronomers by Ruan Yuan and evidential scholars circa 1800.

An article by Alexander Williamson (1829–1890) entitled "Advantage of Science" appeared in the sixth issue in June 1857 and equated national strength with science. Williamson noted that the Chinese were very skilled, but he was critical of them for wasting themselves on civil examination essays (this was to become a common theme among missionaries and reformers). He also disdained Chinese literature and poetry. The first to proclaim that China required scientists and science schools to revive the country, Williamson was also author of most of the theological pieces in the *Shanghae Serial.*

By appealing to political and economic power, however, Williamson and the LMS missionaries were treading a treacherous path, both for them as Christians preaching the gospel of science and for the Chinese who sought military power in that gospel. Like the Jesuits, who failed to convince Ming literati that their *scientia* was inherently Christian, so too the Protestants would find that Chinese reformers would disentangle their rhetoric of science and state power from Protestant theology.

Moreover, the link between science and Christianity that Wylie championed perplexed the LMS. Consequently, the *Shanghae Serial* was stopped when the LMS decided to refocus Inkstone Press on missionary activities. This tension between science and religion in the Protestant movement in China mirrored earlier divisions in the Catholic mission during the late Ming. Dissenting orders had convinced the pope that the Jesuit focus on mathematical astronomy had compromised the transmission of true Christianity.

Despite the end of *Shanghae Serial,* which had been the first publication to use contemporary terms for science, Inkstone Press was the first to introduce modern science in China beyond the treaty ports, although its work began in Shanghai, the most vital of these five important ports. Through the collaboration of Li Shanlan, Xu Shou, Wang Tao, and Hua Hengfang, these new trends influenced the late Qing reformist initiatives of Zeng Guofan (1811–1872) and Li Hongzhang (1823–1901).

Introduction of Modern Mathematics and the Calculus

While collaborating on the *Shanghae Serial,* Li Shanlan and Alexander Wylie also translated a work that they entitled *Step by Step in Algebra and the Differential and Integral Calculus* (1859), which reintroduced the Cartesian algebraic symbols rejected in 1712 by the Kangxi emperor. More importantly, this path-breaking translation of the work by Elias Loomis (1811–1889) entitled *Elements of Analytical Geometry and of the Differential and Integral Calculus,* published in 1851, presented an important improvement that the Jesuits had failed to transmit to China.

Wylie noted in his English preface the effect the calculus would have: "There is little doubt that this branch of the science will commend itself to native mathematicians, in consideration of its obvious utility . . . A spirit of inquiry is abroad among the Chinese, and there is a class of students in the empire, by no means small in number, who receive with avidity instruction on scientific matters from the West. Mere superficial essays and popular digests are far from adequate to satisfy such applicants."[40]

Wylie failed to clarify the historical context behind the historic translation, however. Cambridge reformers in the 1820s had contended that because they used Newtonian fluxions, which obscure the coherence of the calculus, the British had been losing ground in mathematics since the 1740s. Indeed, it took another century for them to catch up to the French, who, using Leibniz's notational forms, had successfully developed a series of rule-like differential and integral equations that engineers did not need to reconceptualize each time.[41] In the nineteenth century, then, De Morgan was committed to French calculus and the elimination of Newtonian fluxions. Similarly, William Whewell's mechanics built on Lagrange's *Mécanique Analytique*, even as he preserved Newton's geometric proofs in the *Principia*.[42]

Thus any analysis of Chinese efforts to master modern Western science should consider the variations in European stages of scientific learning country by country. The predicament the British faced in the 1820s was repeated in the 1850s when Chinese mathematicians such as Li Shanlan and Hua Hengfang worked with Wylie and Fryer to translate the calculus into Chinese.[43] Hence, while Wylie was editing and publishing the *Shanghae Serial* in 1857–1858, he and Li Shanlan were working on translations of several English mathematical works and carrying out sophisticated research into the strengths and limits of native mathematics. Later translators of Western mathematics into Chinese, such as John Fryer, did not appreciate the important role of traditional Chinese mathematics.

In 1874, for example, John Fryer and Hua Hengfang together completed the *Origins of the Differential and Integral Calculus* for the Translation Department in the Jiangnan Arsenal. They translated this version from an 1810 article on Newton's fluxions by William Wallace (1768–1843), which the *Encyclopaedia Britannica* published posthumously in 1853. Fryer's use in 1874 of Wallace's 1810 article, however, meant that as an Englishman he was still loyal to old-fashioned British efforts to maintain Newton's fluxions, however modified, rather than the continental calculus produced at Cambridge and introduced to China by Wylie and Li Shanlan's translation of Loomis's more up-to-date textbook. Neither Li nor Hua likely knew about the British-French rivalries that affected Fryer's introduction of fluxions.[44]

There were three potential groups of Chinese readers of the calculus after 1865: scholar-literati, technicians needing Western mathematics for work in arsenals and state schools, and students at various missionary schools. The major audience for Fryer's and Hua's British version of the calculus was in the arsenals and technical schools, particularly the Jiangnan Arsenal in Shanghai. In the Fuzhou Navy Yard, however, the French engineering influence was paramount.[45]

Wylie and Li ingeniously translated mathematical notations, concepts, and theorems for Loomis's calculus into symbols that looked Chinese enough to be compatible with native "four unknowns" notation. Traditional mathematics thus contributed to shaping the translation of the calculus. The translations for the differential and integral calculus also reflected Li Shanlan's and Hua Hengfang's training in traditional mathematics. Li Shanlan invented the Chinese notations that would become standard for these terms. In the early twentieth century, these traditional forms were replaced by Japanese expressions, ironically after Li Shanlan's and Hua Hengfang's notations were first introduced in Japan.[46]

The Wylie-Li and Fryer-Hua translations emphasized the calculus as a powerful tool to solve dynamic mathematical problems. This approach enabled mathematicians to deal with certain curves, surfaces derived from curves, and solids derived from surfaces, which they could consider accurately only when they related them to motion. Despite the adaptation of Chinese terminology, the introduction of the calculus shocked Chinese literati interested in mathematics because traditional mathematics had nothing similar. Initially, the power of the calculus was seen as a practical technique for calculations, which paralleled its use in Europe by engineers. It did not become a separate field of mathematical theory in China until A. P. Parker's new translation of Loomis's work appeared in 1905. At that time, some tried to show that many essential ideas of the calculus came from the mathematical classics by building on earlier claims about the Chinese origins of algebra.

The initial science and natural history translations by Inkstone

Press and the LMS group of Chinese collaborators represented the work of a small but promising scholarly community of scientific educators in the treaty ports in the 1850s and 1860s. Due to civil war, however, such new developments had little influence in the areas of South China affected by the Taiping Rebellion until Chinese armies loyal to the Qing dynasty recaptured Nanjing in 1864. The deaths during the rebellion of several traditional mathematicians, who had played a key role in reviving traditional algebraic techniques, also weakened the literati community in the Yangzi delta—a community that had long championed the link between evidential studies and mathematics.

After Qing armies defeated the Taipings, the calculus spread through the new arsenals first established in the 1860s. The practical applications of the differential and integral calculus for arms manufacture and ship-building made it essential for Chinese working in the arsenals' new technical schools under British or French advisers. A new understanding of European mathematics among Chinese literati emerged that emphasized the integration of Chinese and Western mathematics. This emerging mixture no longer favored Chinese origins. Until the nineteenth century, literati had understood mathematics as a tool. With the introduction of advanced algebra and calculus after 1850, the Chinese began to view mathematics as a field of learning with its own principles.[47]

The Shanghai Polytechnic and Reading Room

After defeating the Taipings, the weakened Qing dynasty, its Manchu elites, and its literati officials faced up to the new technological requirements for surviving in a world dominated by industrializing nations. Literati who had worked with the Protestant missions in the 1850s began in the 1860s to move from their LMS positions to teaching and translating the new natural sciences for the government's westernizing institutions such as the Jiangnan Arsenal in Shanghai, building bridges between industrial revolution science and traditional Chinese natural studies.

In a March 12, 1874, letter to the *North-China Herald,* the British consul in Shanghai, Walter Medhurst (1823–1885), urged the foreign community to foster "appreciation for all kinds of Western knowledge" and "encourage an interest in polytechnic and general information, rather than in mere literary research." Medhurst later submitted a detailed plan for a "Chinese Polytechnic Institution and Reading Room." It would house Chinese periodicals, newspapers, and translations of standard foreign works. Regular lectures and exhibitions would be organized to promote Western manufactures, a model based on similar exhibitions then prominent in Europe and the United States.[48]

At the organizational meeting on March 24, 1874, Alexander Wylie drew up a list of appropriate books and periodicals. Any definitive list would have been somewhat controversial: at that time, more secular Europeans wished to privilege science over religion, while others still clung to the Protestant agenda for linking Christianity to science. The debate reflected the increasing secularization of scientific training in Europe and the United States after the Darwinian revolution had challenged the natural theology of prominent Protestants. For example, Charles Lyell's *Elements of Geology,* translated by Macgowen and Hua Hengfang at the Jiangnan Arsenal in 1873, was premised on Lyell's determination to refute the biblical chronology of creation.[49]

Generous donations from Chinese high officials in the Yangzi delta and the capital region allowed the Polytechnic Institute to open formally on June 22, 1876. Xu Shou proved an able fundraiser when deficits in the Polytechnic's budget deepened in 1877, but Western subscribers attributed its failure to draw Chinese to its reading room to Xu and his bureaucratic ties. They compared the failure of the Polytechnic in Shanghai to the success of the Hong Kong Museum and attributed the difference to the lack of a tradition of public reading rooms in China.[50]

Indeed, Xu Shou treated the Polytechnic like his private library. The directors noted that Xu and his son, ignoring the wishes of the foreign committee, did not encourage visitors. After Xu Shou's death in 1884, the foreign committee led by John Fryer reasserted control and in 1886 the reading room and exhibition rooms were back in working order. Wang Tao, a veteran translator since the 1850s, took over Xu Shou's responsibilities, and in 1886 Fryer invited Wang to serve as the curator of the Polytechnic and its reading room.

When the Chinese envoy to Germany from 1877 to 1878 visited the academy, however, he criticized what he considered its distasteful practical focus and wrote that he thought it ought to be called the "Hall of Artisans" rather than a high-minded "Academy for the Investigation and Extension of Knowledge." He insinuated that the education there was unsuitable for literati elites, for whom classical studies for the civil examinations remained a superior calling. Fryer's goal of popularizing the sciences to address "the practical requirements and affairs of an ordinary Chinaman's every day life" ignored the gulf between ordinary Chinese and literati. The literati would no more automatically take to the gospel of science than would Oxbridge gentlemen.[51]

In 1885, classes and public lectures were finally implemented, and a scientific essay contest was proposed. Instructional plans included the appointment of a foreign professorship of science, which did not materialize. By 1894, the directors had authorized a program of free lectures on scientific and technological subjects in Chinese, which was tied to the six fields detailed in the Western curriculum of the Polytechnic: mining, electricity, surveying, construction engineering, steam engines, and manufacturing.[52]

The First Scientific Journals

Because the Polytechnic did not attract the interest that the oversight committee in Shanghai had hoped for, Fryer and Xu also created the earliest scientific journal to reach Chinese in the treaty ports. Indirectly affiliated with the school, the journal was known in English initially in 1876 as the *Chinese Scientific Magazine* and later in 1877 as the *Chinese Scientific and Industrial Magazine.* In Chinese the name meant "Compendium for Investigating Things and Extending Knowledge" (GZHB). It first ran monthly issues in 1876–1877 and 1880–1881 in Shanghai before turning into a quarterly from 1890 to 1892.

The new journal drew immediate support from Beijing's Society for the Diffusion of Useful Knowledge in China, which was closing down in 1875 and ending its illustrated *Peking Magazine*—a journal that had been edited by the American William Martin (1827–1916) and Joseph Edkins, among others. Printed since 1872 as a monthly, and supported by the society, the magazine was devoted to Western and international news but also included articles on astronomy, geography, and general science. Martin had worked on the journal while heading and teaching at the Beijing School of Foreign Languages. When publication was stopped, the society's members in Beijing transferred their subscriptions to the Polytechnic and its journal.[53]

Beijing remained central for missionaries because of the official School of Foreign Languages established by the Qing dynasty in 1861, which Martin called the "University of Peking." Qing officials, foreign scholars, and Chinese students at the school promoted foreign affairs in Beijing in the 1860s and early 1870s. In addition, the school's two-tier system of five- and eight-year programs of study, established in 1870, prioritized engineering in the longer program and science and mathematics in the shorter course. The School of Foreign Languages used the *Peking Magazine* to promote science and missionary concerns. In turn, the missionaries supported Li Hongzhang and the political partisans of the post-Taiping Self-Strengthening movement in their efforts to reform the Qing regime.

The journal's free monthlies promoting science and technology were usually reprinted in the *Review of the Times,* which was edited by Young J. Allen (1836–1907) and published weekly in Beijing from 1874 and monthly after 1889. Both the *Magazine* and the *Review* addressed issues that concerned those in the emerging arsenals and shipyards. The topics included astronomy, geography, physics, chemistry, medicine, as well as the technology of railroads, mining, and the telegraph. Biographies of Western scientists were added, and many other articles dealt with political economy and current affairs.

Martin's and Edkins's efforts to use the *Peking Magazine* to promote science in Beijing complemented the numerous contributions that Li Shanlan and his mathematics students made to the journal. Often Li's students had their homework or examination papers in the sciences and the mathematics included in the *Peking Magazine* as a reminder that it represented the School of Foreign Languages. They often replied to "Difficult Questions" and established the precedent for "Answers to Readers' Queries," which was later made a regular feature of the *Chinese Scientific and Industrial Magazine,* Fryer's scientific journal that became the public voice simultaneously of the Shanghai Polytechnic and the Jiangnan Arsenal.

Fryer set up the Chinese Scientific Book Depot in 1884 as a retail outlet for Jiangnan Arsenal and outside translations, which were distributed throughout China by *Shanghai Journal (Shen bao)* published by Ernest Major (1842–1908). Copies of the Polytechnic's journal were available first at twenty-four and then twenty-seven of the most important trading centers in East China and Japan. The number of sales agents increased from thirty in early 1880 to seventy by the end of the year. The Polytechnic's journal went beyond the *Peking Magazine* by focusing on the natural sciences and technology in Europe and the United States. With more Chinese participants in its production, the translations in the Polytechnic's journal improved. It was regularly mentioned in treaty port newspapers such as the influential *Shanghai Journal*, whose empirewide Chinese audience allowed Fryer to reach a broader audience outside the treaty ports.[54]

The journal initially printed three thousand copies and usually sold out in several months. Nine months later, the first nine issues were reprinted in a second edition to meet the demand. With four thousand copies printed per issue at its peak in the 1880s and 1890s, it reached some two thousand readers in the treaty ports. Fryer hoped that popular presentations of mathematics and the industrial sciences would become acceptable among literati and merchants. He expected that the *Chinese Scientific and Industrial Magazine* would also compensate for the limited scope of the Jiangnan Arsenal's translations, which were usually printed in runs of only a few hundred copies.

Later in 1891, reprints of previous issues were sold. Missionary reviewers in China effusively praised the civilizing mission of the issues. A review in the winter 1891 issue described the contents, ranging from photography to steam engines and mathematics, and concluded: "Priestcraft and superstition cannot long maintain their hold under the search-light of modern progress."[55]

After 1880, the magazine shifted its emphasis from introductory essays on science to accounts of its specialized fields. Fryer, for his part, increasingly paid attention to mathematics as the foundation of scientific knowledge, which contributed to the strengthening of the teaching of mathematics and science at the Polytechnic after 1885. According to a General Affairs Office memorial to the emperor of May 18, 1887, which advocated modifying the civil examinations to allow candidates to be examined in mathematics, the

Shanghai Polytechnic was training most of the top students of mathematics in the empire.[56]

The *Chinese Scientific and Industrial Magazine* also published some 317 inquiries as "Letters to the Editor." The letters stressed the practical value of technology and showed less interest in pure science. The queries paralleled the technological interests of those involved in the Self-Strengthening Movement. The large volume of letters to Fryer anticipated a more widespread awakening of interest in science in the 1890s.[57]

Although publication of new articles stopped in 1892, portions of the *Chinese Scientific and Industrial Magazine* were reprinted in 1893, 1896, and 1897. In the 1890s the first four volumes were reissued to meet demand, and sold very well in the Chinese Scientific Book Depot after the Sino-Japanese War in 1895. Later, in 1901–1902, the past issues of the journal were reorganized topically, edited under the title *Collectanea of Science,* and reprinted in Shanghai.

After 1904, the Polytechnic gave up its teaching program, but the library and reading room were reopened on March 6, 1905. By 1906–1907, the library had some fifty thousand volumes. In 1911, the Polytechnic offered to lease the property to the Shanghai municipality to build a secondary school. The new polytechnic public school that opened there in 1917 survives today.[58]

The Chinese Scientific and Industrial Magazine

Fryer presented the *Chinese Scientific and Industrial Magazine* as a Chinese journal within the world community of science. He sent the first issue to the editorial committee of the *Scientific American* with a note that it was his "ambition to make this magazine to China what the *Scientific American* is to every country where the English language is known." Seeing this as a sign of progress and believing that in another generation the West would begin to "feel the industrial competition of China," the editors responded very favorably to Fryer's initiative.[59]

Xu Shou, by contrast, placed the scientific articles within the classical tradition of the investigation of things. Xu's classical gloss for the Chinese name for the Polytechnic prioritized the "extension of knowledge and the investigation of things." In Fryer's English translation, Xu's classical references were flattened into a more literal modern rendering: "Science constitutes the primary basis of personal cultivation and state governing." Xu referred to Fryer as a comprehensive scholar, which was a standard reference to a classical scholar in the late eighteenth century who had incorporated mathematics into his broad learning. Fryer translated this reference to him as "an English learned scholar" who had learned the Chinese language.

Xu then reviewed the new fields of science that Fryer had introduced in his arsenal translations and that would comprise the majority of articles in the *Scientific and Industrial Magazine.* He stressed that the fields of mathematics, such as algebra, calculus, trigonometry, and Newton's mathematical physics, were the foundation for the modern sciences. At the same time, however, Xu emphasized that the new journal would introduce practical learning in metallurgy, geology, coastal defense, navigation, and astronomy.[60]

In addition, every issue during the first years included a final section entitled "Miscellaneous Aspects of Science." This loosely organized part mentioned new theories for explaining earthquakes, for example, and science fairs and industrial exhibitions in Europe and the United States, as well as new products and books. Beginning in 1880, a new illustrated series, in lead articles through spring 1892, was called "Explanations of Scientific Instruments" for scientific and medical applications.

The 1880 issues also began a multipart article entitled "On Chemistry in Public Health." Fryer used this series to present a Euro-American model of "hygienic modernity," which redirected the Chinese medical traditions of "nourishing life" and their individual meditative, sexual, and dietary regimes for enhancing the body toward an abstract modern understanding of the chemical processes of life. In subsequent works on this topic in the 1890s, Fryer added temperance as a precondition for health.[61]

The lead article in the inaugural issue, February 1876, was entitled "Introduction to Modern Science" and appeared serially with copious illustrations in the first twelve issues, ending in January 1877. The series derived from *Chamber's Introduction to Science* and covered major fields or topics of modern science: astronomy, gravitational laws of matter and motion, geology, geography, heat, light, electricity, chemistry, botany, and physical anthropology. Fryer acknowledged that the source was an English primer for children, which contained some three hundred items organized into simple explanations.[62]

In addition to an elementary foray into popular science, the early issues of the *Chinese Scientific and Industrial Magazine* included more technical articles that attacked traditional Chinese medical concepts. In doing so, Fryer was stepping into an area of debate among Chinese scholars. While most literati felt that Western remedies were appropriate for external medicine, Chinese physicians, who were less broadly educated, felt that Western therapies were not as good as native therapeutics for internal medicine. They also believed that Western physicians were as adept as the Chinese in using metals and minerals as medicine, but were less capable in applying herbs and plants. Nevertheless, Western medicine was able to pinpoint the causes of an illness before delineating its cure.[63]

William Muirhead's three-part "Principles of Science," and his "General Discussion of the New Methods of Science" in five parts, addressed the methodology of science. The "Principles of Science" first presented examples based on annual and periodic occurrences in astronomy, physics, and botany. His scientific insight that "all things have an appropriate principle to explain them" resonated with the natural philosophy that literati generally accepted. Muirhead redirected the moral aspects of such natural principles away from the Chinese notion of an uncreated and eternal cosmos, which had bedeviled the Jesuits, to an affirmation of an omniscient and omnipotent prime mover.

In the second part of the essay, Muirhead focused on daily and periodic aspects of the earth's rotation and its effect on plants and living things. Muirhead elaborated this point in 1877, eighteen years after Darwin's *Origin of Species* appeared, to stress an argument by design and prove that a supreme power had created a world of living things appropriate to each habitat. To Muirhead, the daily realities of humans, animals, and plants also encoded a natural way of life informed by a natural theology. Religion, although Muirhead did not mention it, overlapped with science in describing life forms, with religion justifying science. Like the *Shanghae Serial* of the 1850s but less explicitly, the *Chinese Scientific and Industrial Magazine* drew on missionary strategies that had been in use since the Jesuits to highlight Christianity as a theological narrative informing science.[64]

Similarly, Muirhead's series on "General Discussion of the New Methods of Science," first published in April 1877, represented an elaborate, if cursory, history of Western science from Aristotle to Bacon, Newton, and Lavoisier. The narrative was a morality tale explaining how the true principles of nature were discovered. Francis Bacon, for example, had unraveled the mysteries in the natural design of the prime mover. Others had built on Bacon's scientific studies to overcome the Aristotelian natural studies of the Jesuits, just as Copernicus had gainsaid the Ptolemaic geocentric cosmos. Newtonian mechanics and the chemistry of Lavoisier also had made significant strides in grasping the true way of the world, Muirhead added.

Muirhead pointed the Chinese reader to the superior, Protestant provenance of modern science, when compared to its deficient,

Catholic predecessor. In Muirhead's triumphal Anglican pedigree of science, Bacon "was considered the founding father of the experimental sciences." Newton then appeared as an intellectual star whose prodigious discoveries in optics, mechanics, mathematics, and astronomy grew out of Bacon's contributions to scientific methods.

Muirhead did not mention Darwin in any of these articles, nor did he address the implications of natural selection for biological selection among plants or animals. We will see in the next chapter that this strategic silence in the face of the Darwinian brouhaha in England and America was typical.

Literati associated with evidential studies who stressed mathematics and astronomy created the intellectual space needed to legitimate the study of natural studies and mathematics in China after 1860.[65] But Chinese and missionaries stressed different aspects of their joint translations: the literati favored a more native trope for scientific investigation, while the missionaries emphasized the notion of modern science informed by natural theology. Because of this blending of traditional and modern themes, mathematics, chemistry, and other scientific pursuits became acceptable second-tier careers for Chinese literati.

FROM TEXTBOOKS TO DARWIN: MODERN SCIENCE ARRIVES

In the aftermath of the Taiping Rebellion, the Manchu state and its literati elites increasingly welcomed Westerners to help introduce modern science and technology to China. Protestant missionaries responded, seeing science as an opportunity to promote their Christian beliefs through their translations of scientific works into Chinese. Manchu officials and Chinese literati in the late nineteenth century often underestimated the degree to which such translations of Western sciences were imbued with a natural theology—or, as in the case of Darwinism in China, were missing entire sets of new scientific ideas.

Early Science Primers

William Martin's *Elements of Natural Philosophy and Chemistry* (1868) was one of the first texts used widely for science instruction at missionary schools and in the Beijing School of Foreign Languages, where Martin was director. Reprinted in many later editions, the primer covered the fields of hydraulics, meteorology, heat and light, electricity, mechanics, chemistry, and mathematics. The *Guide to Chemistry* by Anatole Billequin (1837–1894) was also popular, and a cursory version of the history of modern science was available to students upon completion of Joseph Edkins's "Outline of New Studies in the Sciences," which was published in the *Chinese Missionary News* from 1872.[1]

Similarly, Alexander Williamson published a series from 1873 to 1876 that he called in English *Natural Theology and the Method of Divine Government* (literally, *Inquiry Based on the Investigation of Things*). The series expanded into a textbook that was used in the Beijing School of Foreign Languages in 1880. Williamson's *Natural Theology* stressed that religion and science were complementary, with "religion as fundamental and science as useful." In the series and book, Williamson tried to demonstrate providential design through science.

In the late 1870s, Fryer and the Translation Department in the Jiangnan Arsenal also emphasized translating a series of introductory textbooks to prepare students for more advanced studies in science and technology. Replacing Hobson's outdated texts, the "Science Outline Series" was accompanied by a "Science Handbook Series." Fryer had traveled to England to gather more materials on science before returning to Shanghai in 1873. He chose the science primers compiled by Henry Roscoe (1833–1915) and others, which represented the current scientific fields in Great Britain. The translation was entitled *Primers for Science Studies*, and the arsenal printed it in 1879–1880 in four volumes dealing with the fields of chemistry, physics, astronomy, and geography.[2]

Macmillan and Company had published Roscoe's *Science Primers* in England in 1872. The series represented a collaboration of several top scholars: Roscoe wrote the volume on chemistry; Sir Archibald Geikie (1835–1914) prepared the volume on physical geography and another on geology; Balfour Stewart (1828–1887), a contemporary of William Thomson (Lord Kelvin, 1824–1907), explained his specialty, thermodynamics and physics; Thomas Huxley (1825–1895) prepared the general introduction; and J. Norman Lockyer (1836–1920) completed the volume on astronomy.

From 1857 to 1870, Roscoe had the support of Huxley, then at Oxford, to remake Owens College (later part of Manchester University) into a scientific college and medical school with a focus on a scientific education. This initiative broke ranks with an Oxbridge education for gentlemen, although Cambridge had had a reputation in mathematics and physics since Newton. Stewart became a professor of natural of philosophy (that is, physics), at Owens.

Roscoe's struggles to make science respectable in England in the 1870s paralleled Fryer's efforts in China. After joining the British Royal Society in 1863, Roscoe was instrumental in organizing the journal *Nature*, which began publishing in November 1869. Although London was the home of the Royal Society, and Manchester had been at the center of the British Industrial Revolution, the gentlemanly practice of science and the grimy production of factories and arsenals had not yet merged enough to encourage an ethos favoring science and technology among both industrialists and men of science.[3]

Such technical fields were also less acceptable socially and intellectually for England's middle class and elites than an Oxbridge classical education. Stewart noted in his physics primer that "thoroughly competent teachers" in scientific subjects "are not yet numerous." Consequently, the late Qing Chinese reaction to modern science should be measured by British elite resistance to a science education, such as the 1880s debate between Huxley and Matthew Arnold (1822–1888) over the significance of science vis-à-vis literature and culture, and not by twenty-first century hindsight.[4]

As a civilian, Roscoe promoted modern science in the 1860s through a public lecture series and by organizing cultural events as part of the Manchester Literary and Philosophical Society. Fryer modeled the science translations undertaken at the Jiangnan Arsenal and the arsenal's secular programs for Shanghai literati and merchants to heighten appreciation for Western learning on Roscoe's activities.[5]

A sample of the twenty-seven works in the "Science Outline Series" published from 1882 to 1898 shows the range and scope of the translation project (Appendix 2). In May 1889, for example, texts for the series were available in algebra, trigonometry, calculus, mensuration, conic sections, drawing and mathematical instruments, and electricity. Nineteen volumes were published from 1882 to 1896; twenty-one volumes appeared between 1886 and 1898. A volume in the "Science Handbook Series" accompanied most of them.[6]

Primers for Science and the Problem of Darwin

By the 1880s, Henry Roscoe's original *Science Primers,* which were limited to just four fields, were out-of-date. While on leave in England, the Qing customs director Robert Hart (1835–1911) discovered an updated and more complete 1880 version of the science series published by Macmillan and Company (London) and by Appleton (New York), which included Huxley's introduction and basic works by Roscoe (chemistry), Geikie (geology and geography), Lockyer (astronomy), and Stewart (physics) that Allen and Zheng Changyan had translated—as well as new works on botany by Joseph Dalton Hooker (1817–1911), physiology by Michael Foster (1836–1907), and logic by W. Stanley Jevons (1835–1882).[7]

Upon his return to China, Hart published the entire series as a new textbook for science instruction in the Beijing School of Foreign Languages and other state schools (Appendix 3). In its summer 1891 issue, the quarterly *Scientific and Industrial Magazine* ballyhooed the publication in 1886 of Joseph Edkins's *Primers of Western Studies.* This new series solicited by Hart was also called *Primers on Science Studies.* Li Hongzhang, then the northern superintendent of trade, actively supported it and wrote an enthusiastic preface. The Imperial Maritime Customs Office published it in Beijing.[8]

When Hart asked Edkins, with the encouragement of Li Hongzhang, to undertake the translation of this more detailed primer series, he was taking an important step in helping Qing officials, with the collaboration of missionaries, to advance science in the dynasty's schools and academies. More than a decade before Huxley's "Evolution and Ethics" appeared in 1898 under its translation by Yan Fu (1853–1921) as *On Evolution,* Edkins introduced Huxley's general views on science, although he modified Huxley's Darwinian views. A noted translator and regular contributor to the *Shanghae Serial,* Edkins had earlier worked in Shanghai with Li Shanlan to translate scientific works that were widely read by those involved in China's Foreign Affairs Movement.[9]

Fields of Modern Science

The opening volume in Edkins's *Primers* presented an "Overview of Western Learning," with the seventh chapter organized around a larger group of twenty-three fields that formally made up the modern sciences (Appendix 4). Each of these categories was initially presented in very general terms to define the scope and delimit the content of the modern sciences for beginners. The theoretical perspective informing this survey and the later specialized volumes was a combination of empiricism based on the experimental method and natural theology. The fields of science were presented in detail, but the complexity of nature was ultimately informed by a prime mover whose plans for creation were illuminated through the study of science. Because things in the world were infinite, they were all conceived by the supreme maker.[10]

The intellectual underpinnings of Edkins's translation were as conservative from the viewpoint of Christianity as they were from the perspective of Qing literati seeking classical parallels to modern science. In neither case was science a revolutionary break with past knowledge. Rather, it was a more precise, cumulative outgrowth of earlier learning, which it supported rather than made obsolete. For both Edkins and his Chinese colleagues, the scientist exhaustively mastered principles through things themselves. Edkins translated Huxley's "Introductory" as the *Primer for the General Learning of Science.* His complete translation of Huxley's work was paralleled by a briefer version others created for the Jiangnan Arsenal Translation Department.[11]

The 1886 edition of the *Primers for Science Studies* was slightly reorganized and later reprinted in Shanghai in 1896 under the title of *Primers of Western Learning.* Edkins's translation of Huxley's "Introductory," included in the 1896 reprint, was also made part of a Shanghai collectanea in 1898. The preface for the 1896 edition noted that the press thought a reprint of Edkins's complete translation was necessary because the original edition had not circulated widely enough before the first Sino-Japanese War. It added that the sciences Edkins introduced were also a valuable boon to literati seeking to master political economy. Because Edkins provided explanatory material, his complete version was more understandable to the Chinese. Significantly, Edkins injected a Christian view of biological evolution into his translations.

Modern Geography and Geology

Volumes three, four, and five of the *Primers* dealt with the earth's topography, physical geography, and geology, respectively, and advanced the literature significantly beyond the late Ming Jesuit maps and eighteenth-century surveys discussed earlier. The new scientific view of the earth emphasized the inner core and outer surface of the planet, and discussions of continents and oceans also superseded local geography, which had marked literati interest in historical geography.

The volume that discussed topography paired complete maps of the physical earth with chapters on the earth's land and sea areas.

The advantages and disadvantages of each ocean and continent were delineated in light of climate, ecology, tides, and temperature. The physical makeup of the earth was presented as a physical explanation of the favorable climates and rivers that made China and India so productive in Asia and why Europe's unique land and ocean orientations had provided it advantages in dominating the globe.[12]

In addition to exploring the oceans' depths and their variable temperature and tides, the primer on topography explained the natural phenomena of volcanoes, the formation of canyons and plateaus, the importance of the continental shelf for rivers, and how sediments were carried by rivers to the deltas and seas. The interaction between continents and the oceans was a major theme for understanding how the physical conditions of the earth had evolved.[13] The second primer addressed the geography of the globe and noted how the earth's position in its solar system and its relations to the moon and sun explained the alternation of day and night and the atmospheric winds. The volume on geology continued this discussion, which examined the earth's crust in light of how stones and water had interacted with plants, some now fossils, to provide a favorable ecology for life on earth.[14]

This account of the inanimate, natural contours of our earth reached the borders between science and natural theology but stopped just short of crossing into the domain of natural theology: instead, geography and geology were presented simply as valuable fields of knowledge that enhanced our ability to live in the world. The accounts of plant, human, and animal life in volumes six through eight of the primers, however, crossed that border.[15]

Botany, Physiology, and Evolution

Edkins strategically placed a natural theology in his translations of botany. Indeed, botanists such as Hooker and physicists such as Stewart had not relinquished notions of a blueprint of nature. In vigorously opposing Darwin, for example, William Thomson maintained that physical laws upheld nature's design. Factually, however, Hooker's analysis of the biology of plants in his primer stressed the importance of chemistry and physics for explaining the makeup of plants and the functioning of their parts. Edkins's translation accordingly accentuated the importance of the chemical aspects of botany.

Earlier in 1858, Edkins, along with Li Shanlan and Alexander Williamson, completed an important translation entitled *Botany*, based on John Lindley's 1849 *Elements of Botany*, which they prepared before the Darwinian challenge. This volume was one of the first to develop Chinese botanical terminology within a modern classificatory framework that went beyond the traditional natural-studies focus on plants as medicine.[16]

To update the evolutionary changes in plants in 1886, Edkins avoided mention of Darwinian natural selection.[17] When we look closely at the original English, however, we find that Edkins massaged Hooker's *Botany* textbook to insinuate a natural theology into what Hooker called the "origin of species," a direct reference to Darwin's work. Edkins presented the notion of origins in Chinese as "why plants that grew on earth were different in kind." Moreover, Hooker had articulated plant differences as a choice between "independent creation" and "evolution," with evolution being a direct reference to Darwin; Edkins refracted this explanation into a choice between times when the creator intervened in creation. In addition, although Edkins translated Hooker's chief reservation about evolution (that the "apparent fixity of a species" survived with only limited changes), he failed to tell his Chinese audience that Hooker, who had no natural theology in his original account, instead praised the theory of evolution and accepted the Darwinian notion that only the "fittest survive."[18]

Edkins's translation did favor an evolutionary perspective to explain the myriad changes in plants over time. Unlike Hooker's discussion, however, his presentation depended on God's creation of life forms. This compromise position in biological evolution resembled the Tychonic system, which had been developed in the seventeenth century as a Catholic compromise between the Ptolemaic and Copernican worldviews. Edkins's translations in the 1886 *Primers* reflected an Anglo-American modification of the Darwinian model into Christian Darwinism. Instead of natural selection, Edkins preferred to define evolution in light of the creator's purposes.

In this manner, Edkins misrepresented Hooker by presenting a Christianized version of the slow, evolutionary changes in plants, which he described as a gradual transformation from original pure forms that the creator had sanctioned to the many forms of plant life today. In addition, he anticipated an objection from someone who might question whether or not plant forms had really changed over hundreds and thousands of years. Going beyond Hooker's circumscribed critique of Darwin, Edkins replied to this query by explaining that because the weather and fertility of the earth have not changed drastically, a new plant would appear very rarely and could not be very different from its immediate predecessor.[19]

Edkins's revision never mentioned Darwin's view of the accumulation of biological changes that enabled some species to compete with other life forms for survival. Instead, the text read as a post-Darwinian natural theology. God remained the ultimate creator of all life forms. Changes over time did occur in the makeup of a particular species, although such changes were always gradual and never drastic. Survival of the fittest in an amoral world of natural selection was as anathema for Protestants as the decentering of the earth had been for Jesuits.[20]

A similar natural theology permeated Edkins's volume on human physiology, which he translated from a work by Michael Foster (1836–1907). Although he detailed Foster's version of the complex process of chemical oxidation that sustained life, Edkins ultimately attributed the physiological processes that an organism used for digestion to the heat the creator had intended for the human body to survive.

Edkins included a theological argument in his description of the importance of oxygen in the bloodstream for energy, which otherwise clearly articulated Foster's summation of human physiology. Human perception of the outer world depended on a remarkable internal network of organs of perception tied to the brain. Within the brain, however, the "soul remained the master of the body," according to Edkins. Again, Edkins's spiritual vision trumped Foster's biological and chemical account of the body.[21]

As up-to-date as the 1886 version of the *Primers* was on the chemical elements, and as pioneering as Edkins's accounts for gases were, Edkins still maintained that the creator had endowed the planet with life and spirit. Theology still took precedence in his account of modern chemistry and physics. Hence, the 1886 version of the *Primers* was a highly mediated mirror of the natural sciences in the aftermath of the Darwinian revolution.[22]

Prize Essay Topics and Their Scientific Content

To attract the interest of the mainstream, Fryer and Wang Tao devised the China Prize Essay Contest in 1886. Trying to tap into the large audience of literati writers and readers who had taken the civil examinations, Fryer conceived of the essay writing contest to attract the many literati proficient in examination essays to write about foreign subjects, including science and technology. Fryer linked the Shanghai Polytechnic's educational focus to the Chinese examination culture around it. In time this experiment became one of the most successful undertakings of Shanghai Polytechnic to spread Western learning beyond the treaty ports.[23]

The Polytechnic's essay contest closely paralleled the civil examination process of reward and fame for prized essays. Mimicking the palace examination, the major prizes were announced in the Chinese and Western press, and the best essays were released to newspapers such as the influential *Shanghai Journal.* Wang Tao also edited special volumes of prize science essays published by the Polytechnic, which paralleled reprinted collections of the civil examination essays. The Qing officials charged with formulating the questions assumed that they were grading the equivalent of policy essays for the palace examination, which normally used traditional styles.[24]

Qing officials quickly saw the efficacy of applying the examination ethos, so well entrenched among the literati, to Western learning. Li Hongzhang and Liu Kunyi (1830–1902) as the northern and southern superintendents of trade, respectively, each consented to give an extra theme every year during the spring and fall contests. Fryer noted that when Li Hongzhang's extra theme for the first half of 1889 was issued, thirty essays were forwarded to him for ranking. Li's list of the twenty-seven successful writers, and the extra awards given them, together with his personal criticisms, were published in the local Chinese papers.

Literati who sent in essays for the competition between 1886 and 1893 were getting their information for the topics mainly from Jiangnan Arsenal translations, materials prepared at the Beijing School of Foreign Languages, and articles on the sciences that had appeared in the *Chinese Scientific and Industrial Magazine,* the *Peking Magazine,* as well as the *Shanghae Serial.* The popularity of the "Answers to Readers' Queries" in the *Peking Magazine* and "Letters to the Editors" in the *Scientific and Industrial Magazine* also indicates the influence these journals had among literati in and outside the treaty ports, although the real boom in reprinting such publications came after China's 1895 defeat in the Sino-Japanese War.

Following these pioneering compilations, other compendia on the sciences became more widely available. They reprinted many science works from the arsenal translations and from Fryer's "Science Outline Series." Literati found out about these new compilations via book advertisements in the emerging Western and modern Chinese press and from catalogs of Western books. Advertisements in the 1897 issues of the *Shanghai Journal*, for instance, included mention of new works on chemistry and public affairs directly aimed at the civil examination market: one advertisement claimed that a particular book on Western history was "absolutely essential for success" on the policy essays required in the civil examinations.[25]

The essay competition was enthusiastically received, and many of the essays were not only printed in newspapers but also included in reformist encyclopedias. For forty-two contests held from 1886 through 1893, in which 2,236 total essays were sent in, an average of 46.1 percent of the essays received prizes. Overall, each contest drew about fifty-three essays. Seventeen Chinese officials presented a total of eighty-six questions, with several presenting questions a number of times.[26]

Altogether, 1,878 Chinese submitted essays over the eight years of the competition (to 1893). Among the ninety-two essayists who were honored (4.9 percent), several produced five or more essays that were awarded prizes. Despite such success and patronage, Fryer was ambivalent: "The fact that only thirty essayists dared to tackle all three subjects is an evidence of the general ignorance of the literati on everything outside the ordinary curriculum of Chinese study; while at the same time it shows how effectively this prize essay scheme is doing its work." Fryer also saw some limits in the sorts of questions that other officials such as the Shanghai circuit intendant or governor-general in Nanjing prepared: "Although their questions relate, perhaps, more to political economy and commerce than to the severer branches of science, it is still gratifying to see how patriotic they are, and how they regard knowledge from the practical, utilitarian point of view, rather than from the theoretical alone."[27]

In time, the Polytechnic prize essays became sources of information, as indicated in Appendix 5 by the publication of the *Compendium of Prize Essays on Science* in 1897 and the *Shanghai Polytechnic Prize Essay Competition* again in 1898. Key bibliographies of Western Learning were also compiled in the 1890s. Xu Bianze's *Books on Eastern and Western Learning*, when published in 1899, praised the *Chinese Scientific and Industrial Magazine* published by the Polytechnic for introducing the sciences to literati since the 1870s.

These essay competitions influenced the reformed civil examinations of 1901–1904, which the court implemented after the antiforeign Boxer Rebellion of 1900 in North China. The reforms required all candidates to master the new fields in the sciences and world affairs. Many candidates who had prepared essays for the prize essay competition, such as Zhong Tianwei (1840–1900), went on to take the reformed civil examinations. In addition, many of the policy questions used in the reformed civil examinations were derived from topics that officials such as Li Hongzhang had chosen for the prize essay competition. When the civil examinations were abrogated in 1904, these scientific texts and science topics remained important in the new schools formed after 1905.

The growth of a significant reformist community of scholar-officials conversant with science, which accompanied the growth of thousands of technicians, engineers, and skilled artisans in the empirewide arsenals, began in the 1880s and 1890s—before the Sino-Japanese War. One such scholar-official was Du Yaquan (1873–1933). Du taught himself mathematics by studying the works of Li Shanlan and Hua Hengfang. In 1898, the future chancellor of Beijing University, Cai Yuanpei (1867–1940), invited him to teach at a new school focusing on Chinese and Western studies in Shaoxing, Zhejiang. Du built on his knowledge of science at the end of the nineteenth century by teaching science and founding the first modern chemistry journal in Shanghai in 1900. In 1904, Du took charge of the Shanghai Commercial Press's science publications section, and in 1911, he became the editor of *Eastern Miscellany,* an influential scholarly journal published by the Shanghai Commercial Press.[28]

The revamping of the civil examinations held for some 150,000 provincial candidates empirewide meant that science questions were now given regularly. For example, five of the eight post-1900 essay topics on the natural sciences were phrased as follows:

1. Much of European science originates from China; we need to stress what became a lost learning as the basis for wealth and power.
2. In the sciences, China and the West are different; use Chinese learning to critique Western learning.
3. Substantiate in detail the theory that Western methods all originate from China.
4. Prove in detail that Western science studies mainly were based on the theories of China's pre-Han masters.
5. Explain why Western science studies are progressively refined and precise.
6. Itemize and demonstrate using commentaries that theories from the Mohist Canon preceded Western theories of mathematical astronomy, optics, and mechanics.

Such questions and their answers revealed that among imperial examiners the wedding of the traditional Chinese sciences and Western science was still intact. The Polytechnic's prize essays thus give us a vantage point from which to evaluate the conventional packaging of modern science among Qing literati (Appendix 5).[29]

For example, Zhong Tianwei prepared an essay for Li Hongzhang's spring 1889 "Extra Theme" on native and Western science. Parts of his policy essay were written following the exact parallelism of an eight-legged civil examination essay:

With regard to the sciences, China and the West are different.
Speaking of it from the angle of what is above form,
then earlier literati scholars clarified everything leaving
nothing out.
Speaking of it from the angle of what is below form,
then the new principles of the West daily emerge without end.
It is likely that
China has emphasized the Way while undervaluing the arts.
Therefore, Chinese science valued meaning and principles.
The West has emphasized the arts and undervalued the Way.
Therefore, Western science has focused on principles of
things.
This is why China and the West diverged.[30]

In a spring 1889 essay prepared for Li Hongzhang's "Extra Theme" on the development of Western science since Aristotle, Wang Zuocai, a student at Shanghai Polytechnic, also argued that "China has stressed the Way and undervalued the arts." He emphasized that since the Greeks, the West had focused on things themselves as objects of analysis in contrast to the political, moral, and institutional focus of classical scholars in imperial China.[31]

Medical Questions as Prize Essay Topics

Since Benjamin Hobson's pioneering translations, Chinese literati and physicians had increasingly noted the anatomical and surgical strengths of Western medicine, when compared to the therapeutic efficacy of classical Chinese medicine. Hobson's work and that of his successors had since the 1850s focused on medical education and hospitals in China. In 1872, for example, the English missionary and physician John Dudgeon introduced courses in anatomy and physiology at the School of Foreign Languages in Beijing. Dudgeon also published a six-volume work in Chinese on anatomy, which the Qing government subsidized in 1887. A companion volume on physiology also appeared.[32]

Between 1874 and 1905, the number of professional medical missionaries rose from ten to about three hundred. In 1876, forty missionary hospitals and dispensaries treated 41,281 patients. Three decades later, 250 treated approximately two million patients annually. Dr. John G. Kerr of the American Presbyterian Mission took over Peter Parker's Guangzhou hospital, and supervised the treatment of over a million patients during his half century of service to his Cantonese patients. Other large missionary hospitals were established in Hangzhou and Tianjin. Li Hongzhang's wife, for example, endowed the Tianjin Hospital to repay Dr. John K. Mackenzie for saving her life.

A generation of Chinese trained as modern physicians also emerged after 1870. By 1897, about three hundred Chinese doctors had graduated from missionary medical schools, with another 250 to 300 in training. Many were trained in missionary hospitals in China or Hong Kong. Numerous fully trained native assistants made up the staffs of most such hospitals and dispensaries. Between 1886 and 1929, some 3,816 Chinese graduated from twenty-four modern private or public medical schools.[33]

Several Chinese revolutionaries studied Western medicine early in their careers. Sun Yat-sen (1866–1925), for example, was a member of the first graduating class of the Hong Kong College of Medicine in 1892. Both Guo Moruo (1892–1978) and Lu Xun (1881–1936) traveled to Japan to study modern medicine, and Guo completed his premedical and medical courses there. Lu Xun turned to literature after the Russo-Japanese War in 1904–1905, but he described his youthful zeal for Western medicine as a reaction against the "unwitting or deliberate charlatans" who posed as classical Chinese physicians. Lu also gleaned from existing translations that the Japanese revival had been based in part on the "introduction of Western medical science to Japan."[34]

Western medicine, not triumphant until the twentieth century, still faced a sizeable opposition among traditional physicians in the late Qing, although many Chinese had joined Xu Shou in his critique of the conceptual weaknesses in Chinese medical theories. Traditional Chinese physicians gradually integrated the Western anatomy of blood vessels and the nervous system. Accordingly, when the Polytechnic included medical topics for the Prize Essay Contest, the rout of classical medicine was still not noticeable.[35]

In spring 1891, for instance, the prize essays appealed to the methods for investigating things that had informed medieval classical learning. In addition, the essays compared the healthful aspects of Chinese and Western foods. Most importantly, however, the essays pointed to the strength of modern chemistry to elucidate the alchemical findings of medieval adepts. Citing Martin's *Elements of Natural Philosophy and Chemistry*, one essay noted that native traditions for "nourishing life" and Western chemistry together could reveal the efficacy of materia medica. When Wang Tao added his evaluations for the published version, he agreed that alchemy had paralleled Western chemistry in important ways.[36]

Similarly, when the official Liu Kunyi prepared an essay topic on medicine for the special fall 1892 Prize Essay Contest, he asked authors to address which medical tradition, Chinese or Western, was superior conceptually. The Zhejiang literatus Xu Keqin, who altogether submitted seven prize essays, stressed the achievements of ancient Chinese physicians, especially Zhang Ji, whose *Treatise on Cold Damage Disorders* had by early Ming times reached canonical status.

Xu added that Western physicians emphasized the nervous system but were unaware of the twelve circulation tracts important to understand the body's susceptibility to illness. In particular, Xu's and the other essays still focused on therapies in Chinese medicine, while pointing to the dangers in the surgical techniques employed by Western physicians. Neither the Chinese nor the missionaries were aware that Patrick Manson was unraveling the natural history of infectious diseases, on the basis of his experiences in Amoy.[37]

In his prize essay for the summer 1893 contest, Xu Keqin addressed a query by the Ningbo circuit intendant Wu Yinsun on the comparative strengths and weaknesses of Chinese and Western medicine. The essayists were asked to demonstrate their knowledge of the history of Western medicine and its most famous physicians. Xu provided such a summary, but he added that Chinese therapies were superior to the invasive, surgical techniques used in Western countries and now in China. He admitted, however, that each tradition had certain strengths that should be selected and combined.

The Anhui literatus Li Jingbang's 1893 prize essay on medicine described the place of the brain in Western medicine. He also noted the institutional importance of the hospital for the success of Western medicine. Li's observation regarding hospitals was interesting because, except for those serving the poor, hospitals were not that prominent by the 1890s. Li claimed that traditional medicine had also stressed the brain, and through his apologetics he tried to attach to medical studies the still prominent claim of the Chinese origins of Western learning.

Li claimed, for example, that during the Roman Empire Westerners had come to China and had taken back the early Chinese canonical medical texts. Over time, Western medicine had built on these texts to produce new findings drawn from chemistry. (We will see in Chapter 7 that such traditionalist claims were increasingly suspect by 1900 as more Chinese were trained as modern physicians.) Moreover, Li blamed the decline of Chinese medicine on its recent practitioners, who had failed to live up to the comprehensive understanding achieved by ancient physicians.

Yang Minhui's prize essay for the summer 1893 medicine theme showed that Western medicine was increasingly respected in official and popular circles and feared by native physicians. Yang tackled this change by appealing to the theoretical superiority of classical medical principles while at the same time admitting the advantages of Western medical practices. Western knowledge of electricity and chemistry, according to Yang, were two areas that when applied therapeutically superseded native medicine. Altogether Yang presented ten areas in which Western medicine was superior, but only four areas in which native medicine prevailed. To salvage the strengths of Chinese medical principles, Yang proposed that the new medical procedures from the West should be used to inform ancient principles.[38]

The Polytechnic's prize essay topics on medicine also influenced the reformed civil examinations after 1901. Essays from both the 1891 Polytechnic query on the medieval materia medica and the 1892 theme "Which medical tradition is superior theoretically?" were presented as model policy essays. Prize essays on materia medica and the history of Western medicine served the same purpose in a 1903 collection of policy answers. More significantly, civil service examiners now regarded medicine as one of the modern sciences.[39]

Natural Theology, Darwin, and Evolution

The extra theme question for the spring 1889 examination, created by Li Hongzhang, asked contestants to explain Darwin's and Spencer's writings (Appendix 5). The essays on this topic reveal more ignorance than knowledge concerning Darwin's explanations about the evolution of life forms. Jiang Tongyin's prize essay, for example, simply bluffed its way through the issues—the judges, including Li Hongzhang, apparently did not know the difference. Jiang's essay received the first prize, but he identified Darwin as a famous geographer who also wrote on chemistry, and described Spencer as an expert in mathematics. Those characterizations may seem odd for someone who came from the outskirts of Shanghai, where Western learning was fairly common, but Jiang himself may have offered an explanation. He added to his essay: "Today the works by both gentlemen are widely prevalent abroad . . . but I regret not yet seeing any translations." Jiang relied on missionary translations, which were silent about Darwin and Spencer.[40]

In his second-place essay, Wang Zuocai, a student at Shanghai Polytechnic, was more discerning. Wang noted that Darwin had argued that organisms adapted to the world in different ways and thus evolved according to different levels of complexity. Without adapting, such life forms could never survive. Although Wang recognized that Darwin had discovered a principle that no one had ever understood before, his account missed the key role of natural selection. Moreover, when he came to Spencer, Wang bluffed his way through by linking Spencer to Darwin and simply stressing the popularity of Darwin's writings.[41]

The third-place finisher, Zhu Dengxu, like the first-place essayist, was totally ignorant of Darwin's and Spencer's views. A supplementary student in Shanghai county, Zhu described Darwin as a geographer who became famous for his travels. Darwin's research, Zhu added, was based on science and chemistry, but Zhu also added a clue explaining his limited knowledge: none of Darwin's major works had been translated into Chinese. According to Zhu, Spencer was skilled in practical applications of mathematics to science.

Such shallow characterizations of Darwin and Spencer in 1889 by Chinese literati in the Shanghai area are at first sight surprising, given John Fryer's ties to British science since his early 1870s home visit. Yet after returning to Shanghai in 1873, Fryer clearly avoided the Darwinian controversy, although favorable mention of Darwin's *The Descent of Man* (1871) had appeared in the *Shanghai Journal* on August 21, 1873. And no works of Darwin or Spencer were translated at the arsenal, although the arsenal did translate a very general piece by Huxley introducing the *Primers for Science* series in 1875.[42]

The silence about Darwin becomes more disquieting when we

examine more carefully the essay that won fourth place, which we have already considered briefly. Zhong Tianwei had received a local degree in the civil examinations at the age of twenty-six *sui* (twenty-five years in Western terms), and he was awaiting appointment as a magistrate's aide in Guangdong province. After failing the more competitive provincial examinations in Nanjing, at around 1873, Zhong, at age thirty-three, entered the Foreign Language School in the Jiangnan Arsenal, where he studied Western learning for three years. Subsequently, Zhong traveled abroad in Europe for two years, then worked with Fryer and others in Jiangnan Arsenal's Translation Department. Because of his ties to the arsenal, both as a student and translator, and his travels abroad, he seems to have had access to accounts of Darwin and Spencer unavailable to others in Shanghai. Zhong's prize essay also drew literally on Edkins's *Primers for Science Studies.*

We have seen that among the works Fryer brought back with him after his leave in England was Roscoe's series of *Science Primers* published in 1872. One volume of the series was written by Thomas Huxley, a champion of Darwin since 1860 when Huxley had debated the bishop of Oxford. Fryer must have heard of Darwin's *Origin of Species* while in England. He had the arsenal's Translation Department translate Roscoe's series, including Huxley's volume. Hence, we find in Zhong's prize essay a remarkably accurate account of Darwin's theory of evolution. Why, then, was it ranked fourth?

The essays submitted for the Chinese Prize Essay Contest contain some of the earliest references in Qing China to Charles Darwin and his theory of evolution. The first, albeit brief, documented reference to Darwin came in the Jiangnan Arsenal translation of Lyell's *Elements of Geology* (sixth edition, 1873) by Hua Hengfang and Macgowan, and again that year in articles in the *Shanghai Journal.*

A vague mention of human evolution also appeared in an anonymous September 1877 article in the *Chinese Scientific and Industrial Magazine* entitled "The Theory of Chaos," in which an anonymous author discussed human evolution from apes in an account of how the world might end. This article argued that it was more important to consider the end of life forms rather than their origin, thereby cleverly sidestepping Darwin. Fryer and his aide usually prepared the anonymous articles in the journal. Jesuits had introduced Copernicus in a similar, misleading manner—though they signed their work—and later were chided by eighteenth-century literati for the contradictions in their presentation.[43]

Zhong Tianwei's 1889 fourth-place essay opened with a well-informed account of Greek science. Then it described the evolution of science from Aristotle to Bacon, Darwin, and Spencer. Zhong, however, went well beyond the natural theology that essays on science in the *Shanghae Serial* and the *Scientific and Industrial Magazine* had added to their translations of botany and biology. After surveying the development of Greek science from natural philosophy to metaphysics and dialectics, Zhong summarized the modern contributions of Francis Bacon, Charles Darwin, and Herbert Spencer:

> Two thousand and three years later, the Englishman Bacon first appeared and transformed Aristotle's theories . . . At the age of thirteen he entered the state school to study, but he dismissed old learning, which demonstrated his independent stance . . . Then he focused on the study of science . . . The main point of his study was that in all scientific matters it was necessary to provide substantiation through demonstrable proof so that each principle can be exhaustively grasped without enunciating that principle a priori. In this manner, through an evidential analysis of the nature of things, its principle will become manifest . . .
>
> Darwin was born in 1809 . . . the grandson of a physician

and the son of a scientist . . . Growing up he was selected to attend Edinburgh University in Scotland. Later he traveled around the globe on an English naval vessel carrying out surveys and preparing drawings while investigating each plant and animal in its ecological setting . . . In 1859 he prepared his magnum opus "on the origin of the species of all things." He also declared the "principle of the survival of the fittest" [literally, "the principle that the strong survive and the weak perish"] in Spencerian terms.

All species of plants and animals undergo changes over time and have never remained unchanged. Those plants and animals that are not successful in adapting slowly perish. Those that successfully adapt survive for the long term. This is the natural principle of the heavenly way [close but not quite "natural selection"]. His theory, however, contradicted the teachings of Jesus, and thus scholars from every country refused to follow his words. At first he was greatly attacked, but today those who honor him have gradually increased. Hence, science underwent a great change, and Darwin can be called a superior man who arises once in a thousand autumns.

As for Herbert Spencer . . . he was eleven years younger than Darwin. His life works mainly expanded on Darwin's theories, enabling people to grasp the principles of psychology . . . He claimed that only the external appearance of all things was knowable. The inner subtleties of all things were in fact unknowable . . . Moreover the changes that all things go through go back in origin to one thing. This one thing is the root, and all other things are its branches.[44]

Although couched in the rhetoric of the investigation of things, Zhong Tianwei's remarkable essay represented a succinct summary of Darwin's theories and introduced Spencer's methodology almost a decade before Yan Fu's Chinese translation of Huxley's lectures on

"Evolution and Ethics" appeared in 1898. Yan, for example, translated natural selection as "heavenly selection," which like the missionary introduction of evolution in botany prepared by Edkins avoided mentioning the key issues of variation and natural selection.

Zhong circumvented the natural theology that informed the missionaries' account of evolution. He noted in his essay that he had seen a recent translation of Spencer's first work, which was called *Essential Guides for Study.* Zhong's student account thus presented Darwin through Spencer's slogan of the survival of the fittest. Zhong tied his account of species variation to the "natural principle of the heavenly way," which contrasted with the more classically domesticated notion of "heavenly selection" that Yan Fu presented a decade later in his translation.[45]

The lower rank that Zhong's essay received, despite its more informed account of Darwin and Spencer, becomes understandable when we review the comments made by Wang Tao, the overall supervisor of the essay contests. In the edition of the 1886–1893 essays that he published as a collection, Wang Tao often included his comments in the top margins of the page. In his comments on Zhong's explication of the principle of the survival of the fittest, Wang wrote of Darwinism: "This essay describes the flourishing of all living things whereby those most suitable survive the longest. What is referred to as 'those most suitable' means 'those most benefited.' The theory that 'under heaven the strong survive and the weak perish' has no basis in fact."[46]

Wang's comments, which exposed Spencer's—not Darwin's—efforts to justify the social order of his time, indicate why the Darwinian view of evolution was unacceptable in the 1880s and 1890s for Protestant missionaries and their converts. Opposition carried over to their longtime collaborators such as Wang Tao, who earlier had worked with the Scottish missionary James Legge to translate the Chinese classics into English. Hence, the lower ranking of

Zhong's 1889 essay indicated the antagonism that Darwin's views provoked among the contest judges, even though Zhong's essay was the only one to describe clearly the controversy of how Darwin's views "contradicted the teachings of Jesus."

Nevertheless, Zhong's essay was awarded fourth place, published in 1889, and reprinted in later collections. As a low-level translator and educator, Zhong supported educational reform in China within the balanced framework of adapting Western science and technology to Chinese learning, thus unifying the practical arts and the Way. His precocious analysis of Darwin revealed the potential for Chinese literati to reject the Christian packaging of the modern sciences, a process that began in earnest after the Sino-Japanese War, at the same time that Zhong continued to award Chinese learning eminence of place in theoretical matters.

Just as the Chinese eventually learned about Copernicus despite the Jesuits' efforts to conceal his theory of a sun-centered cosmos, so too they learned about Darwin in spite of missionary attempts to replace the theory of natural selection with a Christian natural theology. The Chinese Prize Essay Contest reveals that the Protestant enterprise, once its religious agenda was exposed, was no more convincing to many Chinese in the late nineteenth century than the Jesuit translation agenda had been in the seventeenth and eighteenth. Hence, after the Sino-Japanese War, Chinese literati quickly turned to Meiji Japan for the latest currents in modern science.

After Fryer's departure for Berkeley University (later the University of California, Berkeley) in 1896 and Wang Tao's death in 1897, their successors at Shanghai Polytechnic no longer enthusiastically promoted the essay contests, although they were still held in 1901, 1904, 1906, and 1907. We will see in the next chapter that the Sino-Japanese War heightened the disjunction between events before the war and those after. In particular, many of the events from 1865 to 1894 leading up to the establishment of modern science in China were forgotten as the public paid more attention to reformers, iconoclasts, and revolutionaries after the 1895 debacle. Liang Qichao (1873–1929), for instance, disregarded the remarkable expansion of newspapers after 1850 in his own self-serving accounts of a new and critical journalism in the late nineteenth century. In the process, others such as Tan Sitong (1865–1898) and Kang Youwei (1858–1927), who also rose to prominence after the 1894–1895 war, received the credit for many of the contributions that Li Shanlan, Hua Hengfang, Wang Tao, and others such as Xu Shou and Zhong Tianwei had made in breaking new intellectual ground in the 1870s and 1880s. Both Tan Sitong and Kang Youwei as reformers and publicists appropriated science without understanding it fully.[47]

Another example of this displacement was the meteoric rise of Yan Fu as a public figure. His reputation as an iconoclast, the pioneer translator of Spencer, and the introducer of Darwin's theories pushed his earlier career as a Fuzhou Navy Yard schoolteacher and administrator into the background. The spotlight on Yan Fu and others after 1895 has overshadowed the rise of modern science and Western learning in China before the Sino-Japanese War.

GOVERNMENT ARSENALS SPUR NEW TECHNOLOGIES

Although considered marginal because they usually had failed the more prestigious civil examinations, many Chinese literati saw in Western learning and the modern sciences an alternative route to fame and fortune. In addition, Chinese disappointment with Qing military losses convinced many that more radical political, educational, and cultural changes were required to follow Japan's lead in coping with foreign imperialism. Many Chinese elites increasingly opposed the Manchu dynasty in power, and the public quickly repressed the earlier blending of traditional and new technological and scientific learning, which had begun in the 1860s.[1]

Many Protestant missionaries and experts who had aided in the state's scientific projects jumped into the fray and overzealously concluded that the Qing dynasty, the Chinese language, and traditional culture were doomed. In particular, Chinese naval defeats during the late nineteenth century contributed to the transformation of official, elite, and popular perceptions in the Chinese and missionary press, which shaped the emerging national sense of crisis by 1900.[2]

Role Reversal: Missionaries Work for the Dynasty

We have seen that after the treaty ports were opened in 1842, Protestant missionaries quickly established links with literati and common people in their assigned areas, as Catholics had in the seventeenth century. Literati whom the Protestants trained in the sciences began to connect with the ruling dynasty by serving as official advisers and translators after the devastation of the Taiping Rebellion. Many Chinese who had worked for Inkstone Press in Shanghai, for example, moved from the Protestant missions to the dynasty's arsenals and new schools.

Protestant missionaries also worked in the Translation Department of the Jiangnan Arsenal after it was established in Shanghai—much as leading Jesuits had changed their focus from proselytizing among Chinese literati during the late Ming to working in the Qing bureaucracy to gain access to the imperial court. Like the Jesuits, the Protestants remained committed to the gospel of science in China because they thought its success in government would redound to Christianity.

Well-placed Protestants in China moved from their missions in the 1850s and became employees of the Qing dynasty in the 1860s and 1870s. In the 1850s, the LMS hired Xu Shou, Li Shanlan, and Hua Hengfang to work as translators with Wylie and Macgowan at Inkstone Press. And in the 1860s, the Qing government employed Wylie and Macgowan as translators to work with Xu, Li, and Hua in the Qing dynasty's Jiangnan Arsenal. A small coterie of exceptional Chinese literati also joined the translation project as editors and proofreaders. In this milieu, Zhong Tianwei grasped modern evolution long before Yan Fu did in the 1890s, and Zhao Yuanyi (1840–1902), a cousin of Hua Hengfang, became a pioneering translator of Western medical works.[3]

This transition troubled many Protestant missionaries, as it had their Jesuit predecessors, because their medical and scientific work soon outweighed their missionary activities. From the very beginning, Christians faced the quandary of presenting the progressive aspects of Western learning to the Chinese either primarily via medicine and science or chiefly via religious instruction. These endeavors were theoretically complementary, but in practice even one took up a sizable part of a missionary's daily work.[4]

Wylie, for example, worked with Xu Shou to translate Mains's *Manual of the Steam Engine,* which was published by the Jiangnan Arsenal in 1871. Nevertheless, both Wylie and Young J. Allen (1836–1907) returned to missionary work after working in the arsenal for several years because their translation and teaching work for the Qing dynasty had distanced them too much from their calling as missionaries. By contrast, the more secular John Fryer and William Martin, who taught English, political economy, and international law in Beijing from 1864, spent their careers as well-paid and hardworking servants of the Qing state.[5]

During this era, conservative Manchu officials such as Woren (d. 1871), as well as traditionalist literati, echoed earlier critics of the Jesuits and attempted to derail foreign learning in official schools such as the Beijing School of Foreign Languages. Literati who feared that Western learning would subvert state orthodoxy produced several major nineteenth-century anti-Christian tracts, each of which contained substantial sections from early Qing anti-Christian publications. Although never as successful as early Qing xenophobes had been, a few purists affirmed classical learning and ancient models. Reformers neutralized them in the 1870s, however, and they were finally routed in the aftermath of the Sino-Japanese War.[6]

After the Second Opium War, when British and French forces occupied Beijing, resulting in the 1860 Conventions of Beijing, the northern city of Tianjin became a treaty port, and Western embassies and Protestant missions opened in the capital. Foreign armies sacked portions of Beijing, especially the imperial residence and gardens of the Lofty Pavilion in the northwestern suburbs. Under duress, we have seen that the Qing dynasty began an era of reform known as Self-Strengthening, to cope with the external threats it faced from the superior military power of Western armies and navies.[7]

Late Qing Reformers and Science

The direction of science during this period was largely determined by shifts in the political landscape. Manchu leaders of the 1860s reforms had assisted in the negotiations and signed the 1860 Conventions on the behalf of the dynasty. Chinese regional leaders took the lead in the provinces after Nanjing and the south were recovered from Taiping rebels in 1864. The Manchus Prince Gong (Yixin, 1833–1898) and Wenxiang (1818–1876) together led the Grand Council and the newly founded General Affairs Office in 1861. Until Wenxiang died and Prince Gong lost power in the 1870s, these capital-based and provincial groups cooperated under the banner of Self-Strengthening and were largely responsible for hiring the Protestant missionaries to aid the reform efforts.[8]

The 1861 proposal by Prince Gong and Wenxiang to establish the General Affairs Office included a proposal for a School of Foreign Languages in Beijing. Like at the 1712 Kangxi emperor's Academy of Mathematics, students for this school were initially drawn from the eight Manchu banners and not from Han literati. Li Hongzhang advocated similar schools in Guangzhou and Shanghai in 1863 for Chinese bannermen and commoners. His proposal was based on the 1861 recommendation by Feng Guifen (1809–1874) for establishing an arsenal and shipyard in each Chinese port for creating better arms and ships to defend the coast. Feng also stressed establishing schools in Guangzhou and Shanghai for instruction in Western languages.[9]

The dynasty's pursuit of Western technology began in earnest when Zeng Guofan established the Anqing Arsenal in Anhui province in 1862. Two former LMS translators, Xu Shou and Hua Hengfang, served as directors. Yung Wing (Rong Hong, 1828–1912), a Cantonese scholar who graduated from Yale University in 1854, represented Zeng in buying all-purpose machinery in Europe in 1864 (Yung had advised Zeng in 1863 to launch an ironworks in Shanghai). And Li Hongzhang initially established two small arsenals in 1863, the first in Shanghai and another in Songjiang.[10]

The Songjiang Arsenal moved to Suzhou in 1864 when the Qing court purchased a machine shop for China in 1862. Zeng moved the arsenal to Nanjing, where it was renamed the Nanjing Manufacturing Bureau. Although nominally under a Chinese director, in fact Halliday Macartney, a former British army surgeon who had commanded an army against the Taipings, managed the Nanjing Arsenal and produced fuses, shells, friction tubes for firing cannons, and small cannons for the Anhui army. New machinery was added in 1867–1868, along with some British mechanics. By 1869, Nanjing was producing rockets and trying to forge larger guns.[11]

In 1866, the Hunanese general Zuo Zongtang suggested creating a modern navy yard in Fuzhou, Fujian, to build and operate Western-style warships. Prince Gong and the regents of the Tongzhi emperor (r. 1862–1874) quickly authorized the proposal. When Zuo was sent on military campaigns to Chinese Turkestan (Xinjiang) to put down rebellions, Shen Baozhen (1820–1879) became the director-general of the Fuzhou Navy Yard in 1867. By relying on French know-how, Fuzhou quickly became the largest and most modern of all the Chinese military defense industries established in the 1860s and 1870s. It also had the largest gathering of foreign employees. Until the Sino-French War of 1884–1885, Fuzhou remained a major center of French interests.[12]

Subsequently in 1866–1867, the court approved a proposal to add a Department of Mathematics and Astronomy to the Beijing

School of Foreign Languages. The goal was to teach students about modern science through instruction in chemistry, physics, and mechanics. When William Martin returned to Beijing in 1869 to teach physics after a sabbatical in the United States, he assumed the leadership of the Beijing School.

The addition of mathematics and astronomy in particular was vigorously opposed by Woren in an 1866 memorial sent while he was a Hanlin academician and imperial tutor. Although not against the teaching of mathematics per se, Woren was incensed that foreigners such as Martin would be the teachers. When he was asked by the court to recommend native mathematicians and astronomers for service in the new department, however, Woren pleaded for time because those he might appoint were all linked to the missionaries. Subsequently he used an illness as an excuse to avoid an assignment in the General Affairs Office. He was soon relieved of all duties.[13]

Woren's failure to derail the new Department of Mathematics and Astronomy encouraged Chinese literati such as Li Shanlan, who left Shanghai and the Jiangnan Arsenal in 1869 and accepted an appointment as a professor of mathematics in the Beijing School. Li relocated when the government upgraded the School of Foreign Languages to a college and the department of mathematics and astronomy was secured. Li taught mathematics at the school for thirteen years. Although a special civil examination in mathematics was successfully opposed in the 1870s, Li's mathematics examinations at the School of Foreign Languages remained influential.[14]

The Jiangnan Arsenal in Shanghai

In the summer of 1865, Li Hongzhang, then Jiangsu governor, and the Shanghai customs intendant Ding Richang (1823–1882) rented the largest foreign machine shop in China from an American firm in the Shanghai Foreign Settlement. Li later approved purchase of the machine shop and shipyard for the Suzhou Foreign Arms Office. Li's staff imported additional machinery, and subsequently the Qing government established the Jiangnan Machine Manufacturing General Bureau, usually called the Jiangnan Arsenal, to administer the industrial works and educational offices.[15]

These efforts entailed sizable commitments. Initially, the Jiangnan Arsenal used 250,000 taels (the equivalent of 348,000 silver dollars) for production facilities, drawn mainly from maritime customs funds collected at Shanghai. By 1870, the arsenal had moved outside the old city of Shanghai and quickly became the greatest manufacturing center of modern arms in East Asia. Zeng Guofan, Li Hongzhang, and their advisers agreed that the three basic ingredients required for constructing new industries were (1) the manufacture of machines, (2) the creation of a new institutional category of "engineers" (literally, "machine workers"), and (3) the translation of scientific and technical texts. Through the manufacture of armaments, they thought, the Qing state would master contemporary technology and break the Western monopoly of modern warships and cannons.[16]

Technical work at the arsenal was left in the hands of foreigners. Eight of the original machinists were retained, and six hundred workers were transferred in from the American company. By mid-1867, the arsenal was producing fifteen muskets and a hundred twelve-pound shrapnel daily. It produced twelve-pound howitzers at the rate of eighteen per month, which Qing forces used as munitions in the northern Nian wars of the 1860s and to put down Muslim rebellions in the southwest. From 1871 the arsenal produced breech-loading rifles of the Remington type, which were more efficient than muzzle-loading small arms. By the end of 1873, the arsenal had produced 4,200 such rifles, but they were more costly than and inferior to imported Remingtons. In 1874–1875, Li Hongzhang advised establishing a branch to produce powder and cartridges instead.[17]

John Fryer and the Translation Department

In 1867, Xu Shou, Hua Hengfang, and Xu Jianyin initiated a Translation Department at the Jiangnan Arsenal, which Zeng Guofan enlarged in 1868 to include a school to train translators (Figure 6.1). Zeng and Li Hongzhang regarded translation as a foundation for learning the techniques of modern manufacture and the engineering mathematics, that is, the calculus, on which it was based. Their precedents were the late Ming and early Qing translation projects that had successfully reformed the imperial calendar in the Astrocalendrical Bureau by drawing from then-new techniques and models introduced by the Jesuits.[18]

John Fryer worked at the Jiangnan Arsenal after he left the Anglo-Chinese School in Shanghai in 1867, but he was not officially hired until November 1868. Before that, Fryer was professor of English in the School of Foreign Languages in Beijing from 1863 to 1865. When Fryer officially joined the arsenal as a translator of scientific books, he indicated that he welcomed the appointment over teaching, which he did not enjoy, and over missionizing, for which he was considered too secular. A scandal concerning his wife also made his religious ordination impossible.[19]

Others whom the Jiangnan Arsenal hired in the Translation Bureau included Alexander Wylie (who stayed eight years), Daniel Macgowan, and Reverend Carl Kreyer (who stayed for nine years). They were free to choose books for translation without interference from the imperial government. Wylie, for instance, contracted to work with Xu Shou on Mains's *Manual of the Steam Engine.* Fryer worked with Xu Jianyin to translate William Burchett's *Practical Geometry* (1855 edition), and with Li Shanlan he began a translation of Newton's *Principia.* Macgowan, too, joined Hua Hengfang to translate Charles Lyell's (1797–1875) *Elements of Geology,* whose sixth edition had been published in 1873.

By renewing his contract with the Jiangnan Arsenal in 1871 and

Figure 6.1. The Translation Office in the Jiangnan Arsenal: Xu Shou, Hua Hengfang, and Xu Jianyin, ca. 1870s. *Source: Chuanshi yanjiu* 8 (1985).

becoming the head of the new school there, Fryer chose to work for the reform of China. He stayed for twenty-eight years before accepting the Aggasiz Chair in Oriental Languages at Berkeley University and leaving for California in the summer of 1896. Fryer completed a total of 129 translations at the arsenal, with seventy-seven published by the arsenal itself. (Fourteen of those were released between 1896 and 1909 when Fryer was in California.) The missionary-sponsored School and Textbook Committee and the Chinese Scientific Book Depot published the remainder.

Fifty-seven of Fryer's seventy-seven arsenal translations dealt with the natural sciences. His translations of physics were concentrated in the years between 1885 and 1894. He also completed five works on mathematics in 1887–1888 compared to seven from 1871 to 1879. Forty-eight works dealt with applied science, with eighteen on manufacturing. Because the Chinese government prioritized machinery, Fryer early on stressed Chinese adaptations from *The Engineer and Machinist's Drawing Book,* published in Britain in 1855.[20]

Enlightenment through Translation and Terminology

Writing in May 1886 at a symposium on "The Advisability, or the Reverse, of Endeavoring to Convey Western Knowledge to the Chinese through the Medium of Their Own Language," Fryer noted: "Next if we examine Chinese secular literature we find astronomy and mathematics with kindred subjects have always been popular among the Chinese. The most highly prized books on these subjects are the translations or compilations made by the Jesuits two or three centuries ago. These are found in the library of every Chinaman who has any pretensions to general scholarship. Coming down to more recent times . . . [t]here is strong demand for whatever useful knowledge foreigners have to impart. The cry on all sides is for more books."[21] In 1880, Fryer rejected the missionary view that the Chinese language was inadequate for scientific discourse. Moreover, Fryer discarded the notion that English would become a universal language or that China would ever be ruled by foreign powers. He then described the method of translating that he and his co-workers employed and noted the translators' efforts to establish a systematic nomenclature based on three linguistic choices: (1) using existing nomenclature from native works on the arts and sciences, (2) coining new terms by creating a new character, inventing a descriptive term, or phoneticizing a Western term according to the Mandarin dialect, or (3) constructing a general vocabulary list of terms and proper names.

Earlier plans had not been very successful, according to Fryer, because translators had not appreciated the need to use the same terms throughout a series of publications. Fryer noted that Hobson's pioneering terms, for example, had not been followed, which created confusion among Chinese readers. Hence standardized terms became a key goal of the Jiangnan Arsenal's Translation Department. When translating an English work, the foreign employee and Chinese writer collaborated sentence by sentence via dictation in Chinese, which the Chinese colleague then revised. The resulting translation was then printed using traditional woodblocks because of their economy for reprints.[22]

To promote modern science beyond the arsenal, Fryer beginning in 1877 cooperated with other missionaries, such as Calvin Mateer (1836–1908), in the School and Textbook Committee. The committee met regularly in Shanghai to approve textbooks on science and other subjects for use in missionary schools. It planned to prepare a series of elementary and advanced texts covering ten subjects: mathematics, surveying, astronomy, geology, chemistry, zoology, geography, history, language, and music. Although Fryer resigned from the committee for a time, he became the general editor of the series in 1879. At the 1880 meeting, Fryer presented the terms he had used for translation at the Jiangnan Arsenal and compiled a translator's handbook based on his experience.[23]

The need for a consensus on terminology was critical to the success of the science textbooks and primers produced in the 1870s and 1880s. Justus Doolittle's 1872 dictionary of the Chinese language tried to achieve this consensus by incorporating over nine thousand terms from the leading translators of the day for specialized vocabularies in eighty-five fields. Wylie, for instance, prepared a section with 1,016 entries on "Terms Used in Mechanics, with Special Reference to the Steam Engine." Martin presented a list of "Terms Used in Natural Philosophy," and John Kerr (1824–1901) provided chemical terms.[24]

Chemical terminology in particular became contentious among the translators. Considerable variations in Chinese terms for the same chemical elements, for example, occurred in five early and influential chemistry translations (Appendix 6). Several of the translators were then teaching chemistry at the Beijing School of Foreign Languages. On one translation, Fryer and Xu collaborated in the Jiangnan Arsenal's Translation Department, while for another Hobson and Kerr published their translation in Guangzhou. In addition, the German missionary Wilhelm Lobscheid developed an earlier nomenclature from 1866 to 1869 for his English-Chinese dictionary published in Hong Kong, but his translations neglected the phonetic element in Chinese characters and were problematic.[25]

In 1890, Western residents replaced their textbook committee by forming the Educational Association of China, which prepared textbooks for elementary and advanced training in mathematics and science. The association made a concerted effort to unify terminology, but the results were still mixed. By then, it had adopted some thirty-six thousand terms. At the second national Missionary Conference of the Series Committee, which convened in Shanghai in 1890, Fryer reiterated his view that rules for terminology should be strenuously applied. Curiously, he blamed his Chinese collaborators for the inadequacies that others had identified in his translations.[26]

When the 1890 Missionary Committee decided to act, they acknowledged the confusion in the terminology for chemistry. By that time, the Fryer–Xu Shou translation was the most successful in coining Chinese terms for the chemical elements because the collaborators had adopted new Chinese characters using the appropriate signifier, that is, the organizing radical, whereas the Kerr–He Liaoran version had not used fixed principles in their translations. In 1884, moreover, Fryer published through the Jiangnan Arsenal a list of chemical substances based on his *Mirror of the Origins of Chemistry* and its subsequent editions. Fryer's terms for organic and inorganic compounds and chemical concepts, however, were less successful.[27]

Competition among translators was not the only problem. In the case of the Chinese term for mechanics, for example, two legitimate translations competed in the 1860s. The first volume of the *Shanghae Serial* in January 1857 emphasized that its issues would introduce the Newtonian science of mechanics (literally, the "study of weight") to its readers.[28] William Martin, however, challenged this translation in 1868 when he published his *Elements of Natural Philosophy and Chemistry* for instruction at the Beijing Foreign Language School. Martin included a section on mechanics, which he entitled "Introduction to Force." His translation of mechanics as "the study of force" replaced the stress on weight. A similar confusion of terms for mechanics prevailed in Meiji Japan.[29]

Given the potential market for textbooks in the sciences for missionary and dynastic schools and in the arsenal training programs, a missionary-translator could achieve fame and fortune, as Fryer and Martin both did, through the publication of their works. Hence, the competition over translation terminology since the 1850s was also a claim for priority in the lucrative textbook market. The official 1890s response to the continuing controversy over chemical terminology chastised Fryer for not fulfilling his own call for unified rules for nomenclature. Fryer relented, and he became an energetic member of the textbook committee once the controversy had been aired. In 1898 Mateer issued his *Revised List of the Chemical Elements*, which resolved the differences among the three remaining systems of nomenclature developed by Fryer, Kerr, and Billequin.[30]

Fryer also described the enlightenment project of his Translation Department. His goal, which Fryer felt he had in part achieved by 1880, was to break up intellectual stagnation in China. The works the Jiangnan Arsenal had translated were well received, and they were used as textbooks in Beijing at the School of Foreign Languages and in higher-level mission schools, although not by Fryer's rivals in the Jiangnan Arsenal's own Foreign Language School. Moreover, bookstores sold translations to provincial civil-examination candidates in Nanjing. Yet despite the success of his series of specialized and elementary translations, Fryer remained cautionary in 1880 about the popularization of science in China. The "work of supplying useful knowledge to the Chinese by means of their own language" was still needed, Fryer noted, to overturn "the system of ignoring everything but the Four Books and the Five Classics at the Government examinations."

Missionaries working for the Qing dynasty thought they were "bringing about the intellectual and moral regeneration of this great country." Like the Jesuits before them, the Protestants saw science as a way to spread Western knowledge and Christianity. While visiting the Jiangnan Arsenal in 1877, Zeng Guofan presented Fryer with a fan on which Zeng had composed a poem comparing Fryer favorably to Ferdinand Verbiest and Adam Schall. The gift delighted Fryer.[31]

The major limitation in the efforts to unify terminology was the Euro-American deprecation of the contributions made by Chinese collaborators. Moreover, Wylie pointed out in 1867 that because the Jesuits had unwittingly misconstrued many things, particularly in mathematics, the new nomenclature they used for algebra, for example, initially replaced an older established Chinese terminology. There were now two systems of terms, which according to Wylie "have introduced a looseness and inaccuracy of phraseology, little to the advantage of mathematical studies." This looseness continued in the parallel curriculums for modern and traditional mathematics that students in post-Taiping arsenals empirewide tried to master.[32]

Similarly, other terms, such as controlling device *(ji)*, tool or implement *(qi)*, and device or weapon *(xie)*, were applied to technical machinery in the nineteenth century. The technical Chinese name for the Jiangnan Arsenal, for example, was the "Machine Manufacturing General Bureau." When Li Hongzhang proposed establishing new categories for the civil examinations in 1867, he included "mathematics and science" and "technology and manufacturing" as two of the eight categories.[33]

After the Protestant influence peaked in the 1890s, Chinese translators took matters into their own hands and independently translated works in the natural and social sciences from Japanese into Chinese. Generally unaffected by the natural theology encoded in Christian-inspired textbooks, the Japanese science textbooks became models after the Sino-Japanese War. Thereafter, the Chinese translators no longer relied on missionary informants. In chemistry, for instance, Japanese books on the subject continued to use earlier Chinese terms for the elements and compounds, but the terms and concepts informing chemical theory were transformed. The same terminological changes occurred in Japanese books on physics, biology, and geology—and were replicated in the Chinese translations.[34]

Technical Training at Jiangnan and Fuzhou

Before Fryer joined it, the translation project for the Jiangnan Arsenal was very modest. The Chinese and their collaborators had planned to produce an encyclopedia that would resemble the *Encyclopaedia Britannica*, but this goal was too elementary. Instead, the core group of Chinese and Western translators in the Translation

Bureau began producing a series of industrial treatises focused on technology and machinery rather than mathematics and the natural sciences.[35]

From 1863, when the Imperial Court approved its creation, the Shanghai School of Foreign Languages remained independent. In 1869, however, the school was moved to the Jiangnan Arsenal and renamed the School for the Diffusion of Languages. The Shanghai Maritime Customs Department paid for its new buildings. Fryer's work was narrowly defined to translate Western books on manufacturing for the new arsenal school. He initially focused on the fields of engineering, navigation, military technology, and naval affairs.[36]

Teachers in the Foreign Language School also stressed study of the traditional Four Books and Five Classics as part of the lower-division curriculum in the hope that the graduates would pass the more prestigious civil examinations. Hence the school attracted the sons of Shanghai merchants and Christian converts in a more foreign environment, who saw the new learning as an alternate means to access the civil service. Teachers drilled students on the eight-legged essay. At the same time, they used the Ten Mathematical Classics along with the "four unknowns" notational form to tutor the students in traditional Chinese mathematics.[37]

Following the Anglo-American model for training engineers, lower-division students also studied Western algebra, geometry, trigonometry, astronomy, and mechanics. Teachers also provided training in international law, geography, and mechanical drawing. The upper-division curriculum for students emphasized seven fields:

1. Mineralogy and metallurgy
2. Metal casting and forging
3. Wood and iron manufacturing
4. Machinery design and operation
5. Navigation

6. Naval and land warfare
7. Foreign languages, customs, and institutions

It took students three years to complete the two divisions. Outstanding graduates would then take special provincial examinations in Beijing.[38]

At its peak, the Jiangnan Arsenal contained four institutions: the Translation Department, the Foreign Language School, the school for training skilled workmen, and the machine shop. In addition, the Jiangnan Arsenal had thirteen branch factories. By 1892, it occupied seventy-three acres of land, with 1,974 workshops and a total of 2,982 workers. The arsenal acquired 1,037 sets of machines and produced forty-seven kinds of machinery under the supervision of foreign technicians.[39]

Shipbuilding in the Jiangnan Arsenal

From 1868 to 1876, shipbuilding in the Jiangnan Arsenal was highly productive. Eleven ships were built there in eight years. Ten were warships, and of these, five had wooden hulls; the other five, iron hulls. All parts of each ship, including the engine, were built at the arsenal. The arsenal also experimented with different designs, from single to double-screw engines, wooden and iron hulls, and simple warships to turreted vessels. When compared to the warships built following French models at the leading Japanese dockyard in Yokosuka in the 1870s, the level of shipbuilding technology at the Jiangnan Arsenal was actually higher. The Yokosuka Dockyard did not produce its largest wooden warships until 1887–1888, and none matched the largest warship built at the Jiangnan Arsenal in 1872. In addition, the arsenal produced five iron-hulled warships before 1875, while the Japanese did not complete their first iron gunboats until after 1887. In terms of armaments, those manufactured at the Jiangnan Arsenal also were generally superior.[40]

The Chinese fleet of iron and wooden ships quickly fell behind the new ironclad ships of Europe, however. The Chinese began to build compound engines only in 1877. Hence China's ships overall were still behind Europe's in the 1870s. Moreover, because Chinese shipyards could not produce enough ships, more warships were built in Europe for the Chinese navy. Although the Qing continued to employ foreign technicians to build large modern warships in China, Chinese ships remained outmoded by the 1890s because Chinese training did not keep pace with continued Western technological progress. Japanese officers and sailors, in contrast, were better trained to manage their ships and guns by 1894.[41]

Shipbuilding in the Jiangnan Arsenal dramatically slowed after 1876, and in 1885, after the arsenal completed its first steel gunboat, it ceased to be a military shipyard. The technological switch toward steel and armored warships in Europe highlighted the difficulty of transporting iron and coal to make steel in coastal China. At the same time, imported steel remained prohibitively expensive. Nevertheless, shipbuilding technology in Jiangnan and the Fuzhou Navy Yard probably remained slightly better than in Japan until 1889, when a French engineer designed new steel and iron warships for the Yokosuka Dockyard. Its first modern warship had more horsepower and a higher top speed than the same type of warship built at the Jiangnan Arsenal.[42]

Once shipbuilding was no longer its major task, the Jiangnan Arsenal adapted its machinery to produce the most advanced foreign guns and small arms for military use. The goal was to create products similar to those made by the Armstrong factory in Britain. The effort was generally successful: three types of muzzle-loading guns made by the arsenal during this period were deployed at the Wusong fort guarding the mouth of the Yangzi River, and in the late 1880s, the arsenal produced large breech-loading guns that used advanced forms of gunpowder. By 1885, Li Hongzhang was favoring German armaments over the British or French, and the scale of Krupp arms sales to China increased.

Before the Sino-Japanese War, the Jiangnan Arsenal was producing large breech-loading Armstrong guns with a range of 7,000 to 11,000 yards and capable of firing projectiles from eighty to eight hundred pounds. The arsenal also became known after 1890 for its success in producing rapid-firing machine guns, which were important in enhancing sea power and coastal defense forts. Because annual production in the arsenal was insufficient to supply the Chinese army, the Qing military still had to purchase such arms abroad. Japan, by comparison, did not begin its ambitious artillery program until 1905, during the Russo-Japanese War. This program was financed by the sizable Boxer indemnity that Japan received from China for attacks on its embassy in Beijing.[43]

The Fuzhou Navy Yard and French Technology

Besides the Jiangnan Arsenal, the second major industrial site for shipbuilding and training in engineering and technology was the Fuzhou Navy Yard. When Zuo Zongtang submitted his 1866 memorial to establish a complete navy yard at Fuzhou, he expected that after five years he could eliminate the need for foreign experts. The estimated start-up costs of 300,000 taels (the equivalent of 417,000 silver dollars), as well as the 600,000 taels (834,000 silver dollars) for annual operations, came from maritime customs duties and the interprovincial taxes collected in Fujian, Zhejiang, and Guangdong provinces. In return, those provinces would receive naval protection from the Southern Fleet based at Fuzhou.[44]

Zuo and his successor, Shen Baozhen, relied mainly on French expertise for Fuzhou. Once the Qing had established the navy yard, however, the Fujian maritime customs left the venture in a perpetual financial bind. At its peak the shipyard employed 3,000 workers. When later construction was completed the force dropped to 1,900, with 600 in the dockyard, 800 in workshops, and 500 manual laborers. Some 500 soldiers guarded the premises. The Fuzhou Navy

Yard had more than forty-five buildings on 118 acres set aside for administrative, educational, and production purposes. By comparison, the Jiangnan Arsenal, which was China's largest ordnance enterprise in 1875, had thirty-two such buildings on seventy-three acres.[45]

In terms of scale, the Fuzhou Navy Yard was the leading industrial venture in late Qing China. For organizational efficiency, a modern tramway with turntables at important workshops and intersections served the whole plant. Nineteen ships were planned with 80 to 250 horsepower engines. Of these, thirteen would be transport ships with 150 horsepower engines. Sixteen ships were finished during this time. Ten transports with 100 horsepower engines, and one corvette as a showpiece with a 250 horsepower engine, were realized in 1869–1875.

The Fuzhou Navy Yard also compared favorably with the Yokosuka Dockyard. The dockyard at Yokosuka had a budget in 1865 of 1.3 million taels (1.8 million silver dollars) for a four-year period, compared to four million taels (5.6 million silver dollars) allotted to Fuzhou over five years. Actual expenditures at Yokosuka actually doubled the budget, while the Fuzhou Navy Yard expended 5.36 million taels (7.5 million silver dollars) from 1866 to 1874. By 1868, Yokosuka had completed eight ships with eleven more on the way. With two major industrial sites in the Yangzi delta and in Fujian province, the Qing were indeed ahead of Japanese modernization efforts in the 1860s and 1870s. But such aggregate advantages vis-à-vis Japan did not translate into organizational superiority when the Fuzhou naval fleet faced the French flotilla alone and unaided in 1884.[46]

The industrial results in Fuzhou were at first gratifying for the Qing dynasty, and the *North-China Daily News* praised them on December 10, 1875. Ships built at the Jiangnan Arsenal and the Fuzhou Southern Fleet were mainly wooden ships, however, and thus vulnerable to ironclads after the British began producing them in 1865 and the French in 1873. Chinese ships were also not equipped with the latest compound engines. When faced with war with France in the 1880s and Japan in the 1890s, some Qing officials blamed the French, particularly the French naval officer Prosper Giquel (1835–1886), for having dumped obsolete equipment and designs on the Chinese navy.[47]

Giquel had joined the Chinese Imperial Maritime Customs Department as a commissioner of customs and served until 1866. Then he signed a contract to be the foreign director of the Fuzhou Navy Yard. His story is somewhat emblematic of the French influence during the period. At the suggestion of Zuo Zongtang, a school for technical training was opened called the Hall for the Search for Truth, an evidential research slogan, which served as a school for naval administration. Between 1866 and 1874, almost five hundred Chinese students received technical training there according to French standards. Twenty became engineers, and another 348 were highly skilled laborers. An additional thousand men trained as machine and tool workers.

Foreigners taught mathematics, English, French, and drafting at Fuzhou. The Qing dynasty's long-term goal was to mold Chinese naval architects and engineers and to generate modern workmen, carpenters, ironworkers, brass workers, and ship construction workers. In 1873, the navy yard set up two divisions of French and English schools. The French division included departments of naval construction, design, and apprentices. A naval academy with departments of theoretical navigation, practical navigation, and engine room training were in the English division. To train Chinese officers to operate warships, the English division also created a department of theoretical navigation with a mathematical and nautical curriculum that paralleled that for English and French civil engineers. The naval construction department opened in February 1867 with a curriculum that included French, arithmetic, algebra, descriptive and analytic geometry, trigonometry, calculus, physics, and mechanics. The five-year program suffered a high rate of attrition, however. In the first group of 105 beginning students, only thirty-nine remained at the end of 1873.[48]

Besides building the dockyard and training personnel, the navy yard launched fifteen ships between June 1869 and February 1874. But only nineteen were completed between 1874 and 1897 due to the lower caliber of administration after Giquel's departure. The navy yard also faced a decrease in operating funds when Beijing and provincial officials lost interest. Self-management commenced from 1874 when operations in the navy yard were carried on without foreign technicians, until new Frenchmen arrived in 1897.[49]

The schools attracted native students until the late 1880s. Students in the French division were usually from Fujian; those in the English division were from Guangdong or Hong Kong. After 1874 the navy yard sent graduates to Europe, especially England and France, for advanced training. In 1877 Giquel led a party of twenty-six students. Twelve students from the English division went to England, with five of those attending the Royal Naval College at Greenwich. Nine of the fourteen students from the French division studied hull construction and engine principles in France; the other five studied mining and metallurgy. A second group of eight graduates left in late 1882 for three years of advanced training. A third group of thirty-three graduates were sent in 1886, with ten from the English division, fourteen from the French division, and nine from the Tianjin yard. A fourth group was scheduled to go to Europe in 1894, but war with Japan interrupted their departure.

The experience of Yan Fu, a twenty-one-year-old graduate, is illustrative. In 1874 Yan was acting captain of a small steamer owned by the Fujian-Zhejiang administration. As a graduate of the Fuzhou naval division, Yan received advanced training in Europe. Upon his return to China he became a dean and professor of navigation and mathematics for many years at the Fuzhou Navy Yard, and in the early 1880s, the Tianjin Naval Academy appointed him as a professor of navigation and mathematics. He was a teacher and administrator there for nearly twenty years.

After the overwhelming defeat by Japan in the Sino-Japanese War, the navy yard considered an 1896 recommendation to hire foreign teachers in China rather than sending students to European schools, but the Qing court still wished to send the best naval students to Europe for advanced training. Ten were sent in 1897 for six years of schooling. They were recalled in 1900 after three years, however, due to insufficient funds: the huge Boxer indemnities forced upon the Qing state were hampering the work of the empire.[50]

Indeed, industrial decline at the Fuzhou Navy Yard due to financial troubles had set in by 1876–1877. Expenditures exceeded original estimates, partly due to the high monthly wages per person (371.1 silver dollars) for the forty-five foreign advisers and workmen, which used as much as 15 to 24 percent of the overall budget. By contrast, the total wages for two thousand Chinese craftsmen and nine hundred laborers amounted to only 12.5 to 20 percent of total operation costs per month, or 4.8 silver dollars monthly per Chinese worker. Corruption and nepotism ate away at the rest.

The Chinese staff worked with Giquel and his Europeans to keep to the construction schedule, but because the navy yard was financed as a traditional enterprise with numerous sources of income, traditional Qing budgetary practices did not take into account inflation, growth, or retooling. Long-term planning was impossible. After 1880, the Fujian Maritime Customs Department failed to turn over regularly the navy yard's full annual allocation. Consequently, the schools and dockyard became less active in the 1890s.[51]

Naval Warfare and the Scapegoating of Qing Reforms

Europeans and Japanese generally acknowledged that the Jiangnan Arsenal and the Fuzhou Navy Yard were more advanced technically than their chief competitor in Meiji Japan, the Yokosuka Dockyard, until the 1880s. Leaving out the issues of Japanese skills and personal motivations, which were decisive in trumping China's superior numbers in 1895, by the late 1870s China's armaments industries were mainly producing ammunition. Besides financial difficulties, corruption was also rife among leading officials, who competed with each other for the remaining funds.[52]

By 1879, China had two ironclad steamships, which had been ordered from the Vulcan factory in the Baltic for the Northern Fleet and were more advanced than anything the Japanese navy had at the time. (They were both later sunk in the Sino-Japanese War.) In gunpowder manufacture, moreover, the machinery in Shanghai at the Jiangnan Arsenal was more advanced than that used in Germany, although Japan subsequently caught up to China technically in the 1880s.[53]

Disaster in the South

The lack of coordination between the northern and southern navies—not technology—was the chief weakness of the Qing fleets vis-à-vis their counterpart in Japan, which was a unified fleet stationed in Yokosuka. This disadvantage became clearer after 1874 when the French claimed Vietnam as a protectorate, leading to conflict with Qing China in the upper Red River border in northern Vietnam. France also began a naval buildup on the China coast (Map 6.1), which provoked several naval engagements. Although France did not triumph in all the battles of the Sino-French War in 1884–1885, it won the war because of the lack of coordination between the vulnerable Chinese fleet based at Fuzhou and the remote Northern Fleet. The irony that a French-sponsored Chinese navy at anchorage in Fuzhou would be destroyed in a preemptive attack by an invading French flotilla based in Vietnam points to the dangers of relying on European aid in an age of imperialism.[54]

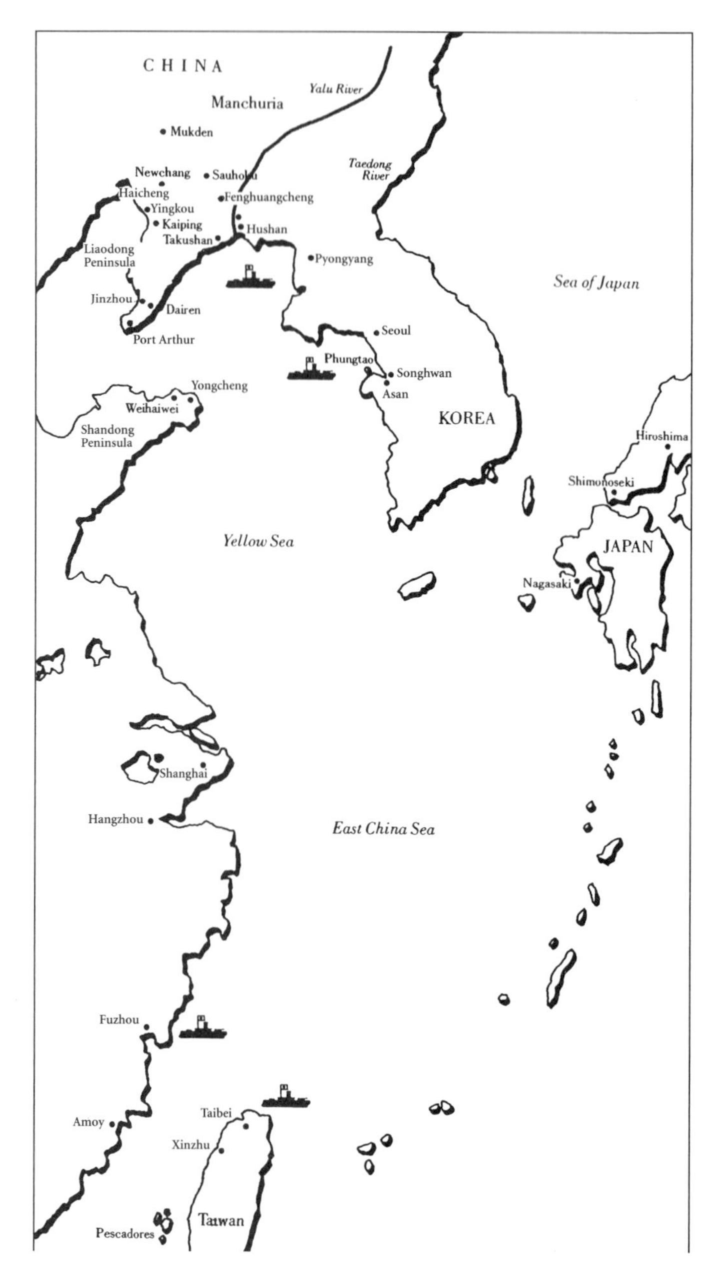

Map 6.1. The Chinese coast during the Sino-French and Sino-Japanese Wars. Map by Phillip Unetic, from Shumpei Okamoto, *Impressions of the Front,* exh. cat., Philadelphia Museum of Art, 1983.

The Qing had over fifty modern naval ships in 1884, with more than half built in China. Its navy, however, was divided into four fleets: the Northern at Weihaiwei and Port Arthur; one in Shanghai; another—known as the Southern Fleet—in Fuzhou; and the smallest in Guangzhou. In the 1884–1885 war, the Fuzhou flotilla fought the French nearly alone in the climactic battle. At its home port of Mawei, the Fuzhou fleet was destroyed in fifteen minutes when French commanders sailed their war vessels uncontested past the river defenses protecting the navy yard. Because the French had not declared war, the French fleet approached the Fuzhou dockyard unchallenged on August 23, 1884. Moreover, before hostilities erupted, Li Hongzhang sent only two of the ships that the Qing court had requested from his Northern fleet, and he quickly withdrew them by asserting that the Japanese threat in Korea mandated their return north.

After the attack, the French fleet withdrew to Taiwan, but after a failed landing French forces threw a blockade around the west coast of the island. Negotiations resumed after a Chinese land victory. China's loss, then, was not simply due to French military superiority: in fact, French technological superiority in the 1880s was not as great as England's during the First and Second Opium Wars, and China had closed the gap somewhat. The actual problems were the political and regional disorganization of the empire, and that naval personnel were insufficiently trained, especially in modern naval strategy.[55]

Except for the lost vessels, the Fuzhou Navy Yard survived with little damage. In the postwar period, however, progress at Fuzhou remained limited. Li Hongzhang thereafter purchased naval vessels for his Northern Fleet rather than building them at home. The Fuzhou Navy Yard also reduced its number of engineers and skilled workmen, although it did launch one ship each year in 1891, 1892, and 1895. The war damaged or destroyed books and logistical supplies for the schools, but officials had restored them by 1886.

The rise of Beiyang as China's chief fleet after 1885 resulted from what Chinese called the "Disaster in the South." Although demanded by the Imperial Court, a single command for a unified naval fleet never materialized. Instead, the new Navy Board and Li's Beiyang fleet competed for financial resources, which were declining due to further naval budget cuts between 1885 and 1894. Efforts by the Empress Dowager Cixi (1835–1908) to garner funds to expand the Summer Palace also drained support from the Chinese navy. Thus inadequate funding and misplaced priorities set limits on Li Hongzhang's expansion of the Northern Fleet.[56]

The Chinese did have some strength in their Beiyang fleet, however, and in the 1880s, after an inconclusive Sino-French War in which the Chinese navy had regained some prestige in the eyes of the Japanese, they showed it off. In the Nagasaki Incident of 1886, for instance, four warships of the Northern Fleet anchored in Nagasaki on their return trip from the Russia. Li Hongzhang sought to impress the Japanese that China's naval equipment, reinforced by new ships purchased from Germany, was superior to Japan's.

China's flaunting of its naval superiority aroused Japanese hostility, eventually provoking the Kobe Incident of 1889, which became a diplomatic dispute after a Chinese port stop there. A reporter described another visit by the Chinese fleet in July 1890, presenting it as an instance of Chinese showing off their new ships. Visitors to the flagship of the Chinese fleet came away impressed with its large caliber guns and thick steel armor.[57]

Disaster in the North

When the Sino-Japanese War unexpectedly began on July 25, 1894, the foreign press generally predicted a Chinese victory. At the time, the Qing navy with sixty-five ships ranked eighth in the world, compared to Japan's thirty-two ships, which ranked eleventh. China's navy was superior in armor, armaments, and tonnage. In fact, some thought that China's two German-built battleships were more powerful than the *Maine* and the *Texas*, the U.S. Navy's largest warships. The vice-admiral in the British Royal Navy believed that the Beiyang fleet in the 1890s was ready for action. On land, however, the sixty battalions of the Chinese army in the north had serious organizational weaknesses—and fewer men. Only twenty thousand front-line troops faced Japan's fifty-thousand-man army.[58]

In Japan, almost fourteen thousand men manned Japan's naval fleet of thirty-two warships and twenty-three torpedo boats. Ten ships were built in Britain, and two in France. The *Yoshino* from Britain's Armstrong shipyard was arguably the fastest vessel of its time when it was timed at twenty-three knots in 1893 trials. China's navy was still divided into four fleets. In 1894, these four combined had about forty-three torpedo boats in addition to sixty-five large ships. The Beiyang fleet more or less equaled Japan's. Chinese ships were equipped with more modern guns, but the navy lacked an adequate supply and transport system to take the offensive. Their defensive posture had contributed to defeat in the Sino-French War a decade earlier.[59]

If foreign opinion in China and Japan favored Li Hongzhang's fleet over Japan's, Japanese newspapers, magazines, and fiction grew excited at the prospect of war with China. Some Japanese were not overly confident of victory, however. The publicist Fukuzawa Yukichi (1835–1901) warned against overconfidence, for instance, although he agreed with Japan's just cause in spreading independence and enlightenment to Korea. The Meiji emperor, for his part, was reluctant to begin hostilities. He had refused to send messengers to the imperial shrines at Ise or to his father's grave to announce the war until he had received news of the initial Japanese victories. Japanese Diet members were also surprised at the easy victory.[60]

Located between Japan and North China, Korea had historically tilted toward China rather than Japan, particularly after Japan's disastrous invasions of the peninsula in the late sixteenth century. Two hundred years later, Korea's Chosŏn rulers still acknowledged their tributary status vis-à-vis the Qing empire. Japan had long been dissatisfied with Qing influence over Korea's foreign policies, and its discontent became a flashpoint between an already contentious Tokyo and Beijing. With the political and economic opening of Korea as the key dispute, hostilities commenced when Japan seized the Korean king. The Korean king's regent, an ally of Japan, then declared war on China. The first encounter between Chinese and Japanese ships occurred on July 25, 1894, at Fengdao, and China's two warships proved no match against an unprovoked attack. After that sea battle, the Qing finally declared war on August 1, sending its Northern Fleet to defend the Chinese coast from Weihaiwei and Port Arthur to the mouth of the Yalu River (Map 6.1).[61]

Subsequently, Japanese military actions at sea and on land stunned the Qing court. The main Beiyang fleet was summoned to the defense: it gathered at the mouth of the Yalu River where a major naval battle with Japan commenced on September 17. It was arguably the first great naval battle between modern steam-powered fleets. Each side had twelve ships in the clash. China had the advantage in armor and weight in a single salvo, while Japan had a decided advantage in speed of ships and total amount of metal thrown in a sustained exchange: it had more quick-firing guns that could fire three times more weight in shells than China's six-inch to twelve-inch caliber guns. While Li Hongzhang stalled and made excuses, the Japanese secured Weihaiwei and Port Arthur, which controlled the entrance to the Bohai Bay and sea approaches to Beijing.[62]

Technology alone was not the key determinant. Japan, for example, could not match China's two major battleships. But Japan proved superior in naval leadership, ship maneuverability, and use of explosive shells. Some observers have rushed to scapegoat the

Fuzhou-trained officers as cowards, believing that because they were the dominant Chinese group and had more experience and training than the Tianjin-trained officers they should have taken the lead. Rawlinson, however, has contended that cowardice was not the decisive factor. China fired 197 twelve-inch projectiles at the decisive naval battle of Yalu, with half of them being only solid shot rather than explosive shell. They scored ten hits, but only four of the successfully aimed projectiles were filled with explosives. In addition, shortages of ammunition hampered the Chinese with their bigger guns. In fact, some of the projectiles (for example, the ones that struck Japan's battleships) were filled with cement, suggesting that there were serious corruption problems in Li Hongzhang's supply command. The Japanese, by contrast, used their rapid-firing guns to score on about 15 percent of their tries. With hindsight, it is clear that the speed and rapidity of fire of Japan's ships were more important at Yalu than the weight of the Qing vessels and their superior armor.[63]

Shore engagements continued after the battle at the Yalu as the Japanese took advantage of their unexpectedly decisive victories at sea to launch a land war, which allowed the Japanese First Army to cross the Yalu and enter China at the Manchurian border. Japanese land forces now had a clear path to march on Beijing. This threat to the capital forced the Qing court to seek an immediate settlement to end the war. The Japanese Second Army, formed in September 1894, also landed on the Liaodong Peninsula and took Port Arthur.

The two main explanations given for China's losses in the land war are similar to those given for the naval war: the better military training of Japanese troops and officers compared to their Chinese counterparts, and that Qing troops were outnumbered by the Japanese at the major battles. A British observer also noted that Chinese crews engaged in the war were at half-strength, but salaries for full crews were paid. In addition, Japanese cryptographers had since June 1894 decoded Li Hongzhang's military communications.[64]

The greatest contrast was in coordination: Japan's navy was unified, whereas in the end the Beiyang navy fought the Japanese principally alone. Li Hongzhang had kept his fleet out of the Fuzhou battle in 1884, and the Nanyang officers now got their revenge by keeping their fleet out of war with Japan. The poor command structure of the Beiyang Fleet and the lack of a court-martial system made it impossible to place blame on any Qing officers or allocate reward properly. Many Chinese captains and officers simply committed suicide. No one dared question the command structure or demand an independent board of review.[65]

The Sino-Japanese War generated intense Japanese self-confidence after 1895. Enhanced by the capture of twelve Chinese warships and seven torpedo boats during hostilities, the Japanese navy added significant tonnage to the Meiji fleet. Moreover, Japanese industrialization and militarization accelerated after the Qing dynasty was forced to pay a considerable indemnity to the Meiji regime. The Japanese government used the 1895 Qing indemnity of 200 million taels of silver (278 million silver dollars) and later Boxer indemnities as windfalls to bankroll a massive rearmament program to address Russian expansion on the borders of northeast China. In the process, Korea and Taiwan were ceded to Japan and became productive colonies.

Meanwhile, however, the Japanese victory had angered the Russians, who feared Japanese expansion on the Asian continent. In concert with Germany and France, the Russians joined in a Triple Intervention after the Treaty of Shimonoseki was signed in April 1895, which forced the Japanese to return the key Liaodong Peninsula to China in exchange for an additional payment from the Qing government. Subsequently in June 1896 Russia and the Qing government signed a secret alliance against Japan in which Russia was granted a railroad concession in the northeastern provinces that were increasingly being referred to as Manchuria. Japan, for its part, allied itself with Great Britain.[66]

The indemnities meant that China's huge payments to Japan (and later, Europe) could not be used to augment the Qing dynasty's reconstruction projects. Qing reparations amounted to 450 million taels (625 million silver dollars) plus interest. This sum was never fully paid, but an estimated 669 million taels (930 million silver dollars) were transferred from China to the foreign countries involved. The Jiangnan Arsenal and Fuzhou Navy Yard in particular never recovered from the indemnities. If the Qing government had been previously unable to integrate development so that innovative institutions reinforced each other, the added weight of Japanese and European imperialism after 1895 meant that the Qing reforms initiated in 1865 had even less of a chance of short-term success.[67]

For the Japanese public, the victory developed into the key event that energized the newly emergent Meiji press (Figure 6.2). Public rage was also directed at the European powers for intervening on the side of China. When Russia later forced the Qing to lease the Liaodong Peninsula to it, the angry Japanese were primed for war with Moscow over the fate of China. Public enthusiasm for military adventures became a common feature after 1895, when dissemination of the national news became a central feature of the Japanese press. There were 600,000 newspaper subscribers in Tokyo and Osaka alone. The naval victory over Russia in 1904–1905 cemented Japan's national exuberance.

In a completely opposite way, the naval disasters and the decisive Qing defeat in the Sino-Japanese War energized criticism by the Chinese public of the dynasty's inadequate policies and enervated the staunch conservatives at court and reformers in the provinces who had opposed Westernization. The unexpected naval disaster at the hands of Japan and the way it was presented as Japan's technological victory shocked many literati and officials, leading to the emergence in literati circles of a greater respect for Western studies. The renewed success of the Shanghai Polytechnic in 1896, for example, was tied to this event. As John Fryer reported: "The book

Figure 6.2. Nakamura Shūkō, *The Great Victory of Japanese Warships off Haiyang Island, 1894.* Meiji era, printed and published October 1, 1894. Reprinted from *Japan at the Dawn of the Modern Age: Woodblock Prints from the Meiji Period,* catalogue by Louise E. Virgin, by permission of the publishers, Museum of Fine Arts, Boston. Photograph © 2004, Museum of Fine Arts, Boston. Anonymous gift RE523288.

business is advancing with rapid strides all over China, and the printers cannot keep pace with it. China is awakening at last."[68]

Unfortunately for Fryer and the missionaries, after the Sino-Japanese War China increasingly imported science books that had been translated or edited in Japan. Accordingly, missionary-based Chinese terminology for science, as used in science journals and textbooks, was soon overtaken by Japan-based Chinese terminology.

Reconsidering the Foreign Affairs Movement

The naval battles that China lost during the late nineteenth century have been held up as demonstrating the failure of the Self-Strengthening reforms initiated after the Taiping Rebellion, as well as the Foreign Affairs Movement. But the rise of the new arsenals, shipyards, technical schools, and translation bureaus, which are usually undervalued in such "failure narratives," should be reconsidered in light of the increased training in military technology and education in Western science available to Chinese after 1865.[69]

In particular, we should not underestimate the significance of the schools and factories launched within the Jiangnan Arsenal in Shanghai and the Fuzhou Navy Yard. John Fryer wrote in 1880 that since 1871 the Jiangnan Arsenal had published some ninety-eight translations of Western works in 235 volumes. Of these, twenty-two dealt with mathematics, fifteen were on naval and military science, and thirteen covered the arts and manufactures. Fryer reported that another forty-five works in 142 volumes were translated but not yet published, and thirteen other works were in process with thirty-four volumes already completed.

Altogether, the Translation Office in the Jiangnan Arsenal had by 1880 sold 31,111 copies, and this had been accomplished without advertisements or postal arrangements. Yet the overall sales were not dramatic. A work on the German Krupp guns translated in

1872, for example, sold 904 copies in eight years. Another work on coastal defense published in 1871 sold 1,114 copies in nine years. *A Treatise on Practical Geometry* (1871) sold a thousand copies in eight years; *A Treatise on Algebra* (1873) sold 781 copies in seven years. Fryer's work on coal mining published in 1871 sold 840 copies in nine years. Publicizing these works beyond Shanghai, Beijing, and the treaty ports was difficult, but even for those venues such numbers were disappointing.[70]

Nevertheless, we should note that Chinese trend-setters were the buyers, and they eventually tilted the balance in favor of science among China's youth. The controversial political reformer cum New Text iconoclast Kang Youwei, for instance, purchased all of the arsenal works when he was in Shanghai in 1882. And between 1890 and 1892, his disciple Liang Qichao purchased many of the arsenal's translations and the *Chinese Scientific Magazine.* Liang developed an influential reading list based on these materials; this "Bibliography of Western Learning" was revised and published in 1896. Of these 329 published works in twenty-eight categories, 119 (36 percent) were translated by Fryer and his Chinese collaborators. The martyred reformer Tan Sitong, meanwhile, wrote in 1894 on scientific topics and mentioned the *Chinese Scientific Magazine* as one of his sources of scientific learning. Tan had visited Shanghai in 1893 and bought many of the arsenal's science translations that had been published by the Society for the Study of National Strengthening.[71]

Besides their use in the missionary schools, the texts guided instruction in a nineteenth-century regional matrix of arsenals, factories, and technical schools that helped spark the twentieth-century industrial revolution in China (Map 6.2). Hence we should also acknowledge the scope and scale of scientific translation and military arsenals elsewhere in China after 1860 (Appendix 7). Not all of them were based on British or French models, although our two examples of the arsenal in Shanghai and navy yard in Fuzhou were.[72]

Once the attack on the Fuzhou Navy Yard during the Sino-

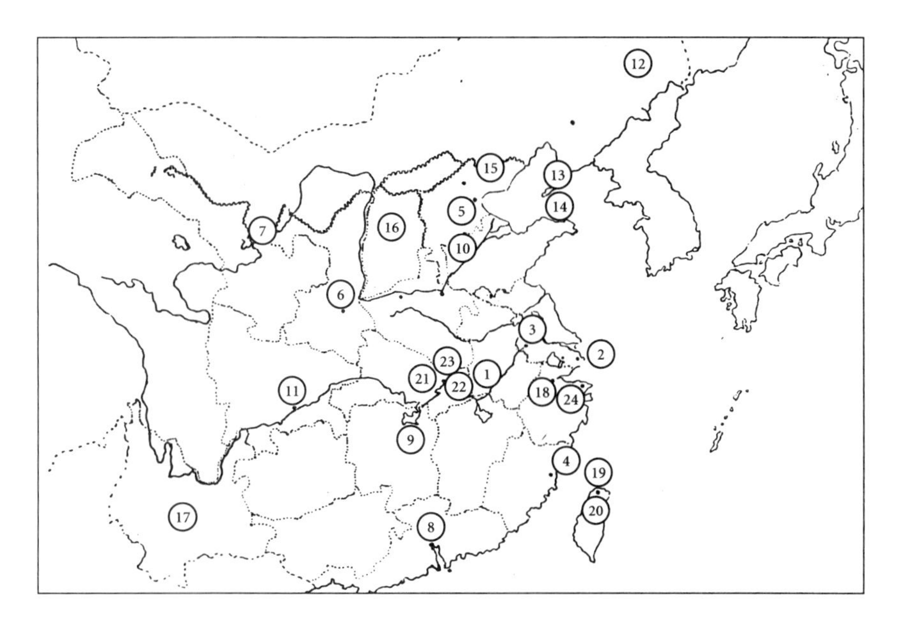

Map 6.2. Partial chronological list of arsenals, etc., in China, 1861–1892.

1. **Anqing Arsenal,** 1861
2. **Jiangnan Arsenal,** 1865
3. **Jinling Arsenal,** 1865
4. **Fuzhou Shipyard,** 1866
5. **Tianjin Arsenal,** 1867
6. **Xi'an Arsenal,** 1869
7. **Lanzhou Arsenal,** 1871–1872
8. **Guangzhou Arsenal,** 1874
9. **Hunan Arsenal,** 1875
10. **Shandong Arsenal,** 1875
11. **Sichuan Arsenal,** 1877
12. **Jilin Arsenal,** 1881
13. **Lüshun, Port Arthur Naval Station,** 1881–1882
14. **Weihaiwei Shipyard,** 1882
15. **Beijing Field Force Arsenal,** 1883
16. **Shanxi Machine Shop,** 1884
17. **Yunnan Arsenal,** 1884
18. **Hangzhou Arsenal,** 1885
19. **Taiwan Machine Shop,** 1885
20. **Taiwan Arsenal,** 1885
21. **Daye Iron Mine,** 1890
22. **Hanyang Ironworks,** 1890
23. **Hanyang Arsenal,** 1892
24. **Zhejiang Machine Shop,** 1893

French War had demonstrated the vulnerability of the coastal factories and fleets to a foreign naval blockade, Zhang Zhidong (1837–1909), then governor-general in the middle Yangzi region, recognized the need for protected inland industrial sites such as the Hanyang Ironworks (1890) and Hanyang Arsenal (1892), both near Hankou. Because his plan was not funded until 1891–1895, however, and then was subject to competing interests of the Northern Fleet and the military threat in Korea, funds were inadequate for simultaneous development of the ironworks and the arsenal. Hanyang failed to produce weapons or ordnance in time for the Sino-Japanese War.

Other delays in plant building and a damaging fire in the summer of 1894 kept the Hanyang project from achieving success before the twentieth century. Zhang wrestled with the twin goals of strategic industrialization and modern military production in the midst of the court's emergency diversion of funds and resources to deal with the Russian and Japanese threats in the northeast. He chose to fund the ironworks for raw material rather than the arsenal for military arms. Hence over the long run the Hanyang Ironworks became the hub of China's iron and steel industry during the first half of the twentieth century.[73]

If we repopulate this impressive list of factories with the scientific careers they fostered, then we can trace more clearly the post-Taiping successors to the native mathematical astronomers that Ruan Yuan's compilation of the *Biographies of Astronomers and Mathematicians* had adumbrated circa 1800. A new group of artisans, technicians, and engineers emerged between 1865 and 1895 whose expertise no longer depended on the fields of classical learning monopolized by the customary scholar-officials. Increasingly, they were no longer beholden to the dynastic orthodoxy or its old-fashioned representatives such as Woren.[74]

The new students of the sciences in the arsenals and missionary schools, who were still a necessary part of the cultural, political, and social hierarchies, emerged from the older categories of myriad elite aspirants for official status. The scientist was "one who investigated things," and he now coexisted with the orthodox classical scholar in the bureaucratic apparatus but at lower levels of political rank, cultural distinction, and social esteem. The self-taught students of modern science and technology in the 1850s, such as Xu Shou, Hua Hengfang, Xu Jianyin, and Li Shanlan, were successors of the Yangzi delta mathematical astronomers who had developed during the rise of mathematics in the age of evidential research. They were in turn succeeded by those like Yan Fu and Lu Xun who were drawn to the Fuzhou Navy Yard and the Jiangnan Arsenal for formal training in science, mathematics, and engineering.[75]

By going outside the orthodox curriculum of the civil service examination, those newly educated in science, mathematics, and engineering inhabited the unprecedented arsenals, shipyards, and factories that offered non-degree-oriented engineering, mathematical, and science training. By linking science and technology, late Qing reformers produced an early version of twentieth-century Chinese techno-science (*keji*). The regional leaders of the Foreign Affairs Movement emphasized technical expertise in engineering and mechanics as well as specialized knowledge of the modern sciences.[76]

Those who were drawn to scholarly work in the new industrial arsenals in Fuzhou, Shanghai, and elsewhere, or translation positions in the three foreign language schools in Beijing, Shanghai, and Guangzhou, still tended to be Manchu bannermen or Han Chinese literati who had failed the civil examinations and saw Western learning and the sciences as an alternative career. Lu Xun and Yan Fu were famous examples of this group of highly educated outsiders from the civil examinations who initially filled the promising late-Qing institutions oriented toward science. Their influence was significant: Kang Youwei and the key reformers were to purchase and read their science textbooks.[77] In Lu's case, a family scandal forced him to leave his lineage school where he had prepared for the civil examinations. He trained at the Jiangnan Arsenal, and later traveled to Japan to study modern medicine at Sendai just before the 1904–1905 Russo-Japanese War, after which he turned to literature.[78] Yan Fu, who was known as a translator and publicist critical of late Qing reform efforts, was as we have seen a graduate of the Fuzhou naval division and later received advanced training in Europe. In 1902 he was appointed chief editor for the new official Translation Bureau in Beijing after achieving fame for his translations of John Stuart Mill's *On Liberty* and Herbert Spencer's social Darwinism.[79]

Eventually thousands of administrative experts, translators, and advisers—including hundreds of foreigners—served in provincial schools and arsenals under the chief provincial ministers of the late Qing. Chinese regional and provincial elites were the tip of an iceberg, the leaders of the post-Taiping turn toward foreign studies focusing on science and industry. Literati associated with statecraft and evidential studies after the Taiping Rebellion legitimated literati study of natural studies and mathematics within the framework of Chinese studies.[80]

The promising start made in missionary schools and the empire-wide arsenals accelerated in the 1880s when Shanghai and Beijing took the lead in promoting the new fields associated with the Foreign Studies Movement. In particular, the shipbuilding industry played an indispensable role in the emergence of late Qing industrial enterprises. China's first lathes and furnaces to produce molten steel were created at Jiangnan and Fuzhou. During the last years of World War I, more than two hundred skilled workers at the Fuzhou Navy Yard manufactured the first Chinese airplane. The arsenals, machine shops, and shipyards provided the institutional venues for an education in science and engineering. They also trained the architects, engineers, and technicians who later provided the manpower for China's increasing number of public and private industries in the early twentieth century.

China's defeats in the Sino-French and Sino-Japanese wars, however, produced a pessimistic intellectual climate. Reformist literati increasingly believed that China was doomed unless the Qing carried out more radical political initiatives. In the process, rhetoric favoring modern science became a key theme of revolutionaries in their political discourses. Earlier efforts to blend traditional Chinese natural studies and mathematics with modern science were set aside in a new perspective that was to inform elite discourse for much of the twentieth century.

CHAPTER SEVEN

THE DISPLACEMENT OF TRADITIONAL CHINESE SCIENCE AND MEDICINE

Despite the relative success of traditional natural studies and modern Western science in developing together in the late nineteenth century, Chinese and their Protestant informants largely ignored the role of laboratories in modern science to discover and test new findings. For Catholic or Protestant missionaries and literati mathematicians, natural studies rarely meant more than translating technical knowledge and memorizing and applying newly available texts. Meanwhile, the techno-science practiced in the arsenals was exceptional, but the practical focus there on producing arms and ships precluded cutting-edge research.

Japan quickly replaced England or France as the nation the Qing dynasty should emulate in science and technology. During this period, for example, the Chinese preferred translating Japanese mathematics texts, which proved to be a convenient shortcut to modern mathematics. The conversion to Western mathematics was also aided by the many Chinese students who returned from studies abroad, particularly from Japan after 1895. Over ten thousand Chinese traveled to Japan to study from 1902 to 1907. Some 90 percent of the foreign-trained students who joined the Qing civil service after 1905, for instance, graduated from Japanese schools. They further tipped the balance toward modern science.[1]

Western Learning Mediated by Japan

Since 1865, literati inside and outside the bureaucracy had distinguished between Chinese learning, which they presented as the whole of native learning, and Western studies. Each was politically charged in the 1890s when they were used by conservatives and radicals in the struggle for or against westernization. When Western learning gained momentum as a model for science and modern institutions after 1895, literati replaced traditional terms for science with Japanese translations. They considered educational institutions that used such accommodations old-fashioned.

The ever-increasing numbers of overseas Chinese students in Japan, Europe, and the United States learned a language of modern science that was not filtered through traditionalist notions. Nevertheless, continuities remained. For example, Japanese scholars during the early Meiji period, influenced by the rise of Germany, demarcated the new sciences by creating a new term for *Wissenschaft* as a broad sense of European science: *kagaku,* which literally means "classified learning based on technical training."[2]

After 1895, Chinese students and scholars adopted both Japanese technical learning and natural studies. Yan Fu, for instance, rendered the sciences using Japanese terminology in his 1900–1902 translation of John Stuart Mill's *System of Logic,* while translating natural philosophy as the investigation of things. Similarly, when regulations for modern schools were promulgated in 1903, the Chinese term for science *(gezhi)* referred collectively to the sciences in general, while the Japanese term *(kagaku)* designated the sciences as individual, technical disciplines. This two-track compromise in terminology lasted through the end of the Qing dynasty and continued after the 1911 revolution. Chinese students who returned from abroad increasingly emphasized a single, modern Japanese term for the Western sciences, abandoning the earlier accommodation between traditional Chinese natural studies and modern science.[3]

Thus the hybrid identity of science and technology as both native and Western declined after 1895. But remnants existed. Although hundreds of Protestant translations printed from 1865 to 1898 had delineated the modern sciences, many conservatives in Qing official circles still asserted the strategic myth that all Western learning could be traced to ancient China. Some late Qing reformers also argued that new political institutions and conceptions from the West were rooted in classical sources.[4] During the seventeenth and eighteenth centuries, traditionalistic rhetoric was plausible because it legitimated the recovery of early Chinese mathematics and ancient learning. After the Sino-Japanese War, however, Chinese self-confidence about their native traditions and institutions dissolved. Such packaging now appeared to be part of the problem rather than an indispensable compromise.[5]

Science and the 1898 Reformers

One of the institutional products of the political iconoclasm in China after the Sino-Japanese War was the Imperial University of Beijing. The Qing government established it at the pinnacle of an empirewide network of schools that would expand on the regional foreign language schools. Like the Translation College, the new university trained those who already had civil degrees in Western subjects suitable for government service. The court again chose William Martin, a distinguished missionary who had worked in the Beijing School, as the dean of the Western studies faculty in 1895.

The curriculum at the Imperial University comprised eight fields: classical studies, politics, literature, medicine, science, agriculture, engineering, and commerce. Six courses defined the science field:

mathematics, astronomy, physics, zoology, botany, and geology. The Imperial University still referred to science courses in light of the investigation of things, although the facilities included modern laboratories equipped with the latest instruments for physics, geometry, and chemistry. This promising development was short-lived, however, because northern rebels associated with the Boxer Rebellion smashed everything in sight at the university in the summer of 1900. European armies were not any kinder during their occupation of Beijing after the Boxer siege of the foreign legations was lifted.[6]

The Qing race to establish Chinese institutions of higher learning that would stress modern science accelerated after Western and Japanese troops occupied the capital in 1900. The Boxer popular rebellion in North China and the response of the Western powers and Japan to it unbalanced the power structure in the capital so much that foreigners were able to put considerable pressure on provincial and metropolitan leaders. Foreign support of reform and Western education thus strengthened the political fortunes of the provincial reformers who had opposed the Boxers.[7]

Westernizing elites who played important roles in the 1898 reform movement enunciated a more assertive approach to modern science. Chen Chi, a secretary in the Bureau of Revenue and a reform advocate, believed that scientific knowledge was a prerequisite for economic productivity in both industry and agriculture. Anticipating the objections of political and cultural conservatives, however, Chen still couched his arguments in favor of science and technology within "Chinese origins" rhetoric. Kang Youwei, like Chen Chi, believed that the military successes of Meiji Japan were a model for China and that emulating the Japanese would require expanded education in the sciences and industry.[8]

In his 1905 essay on industrialization, for instance, Kang emphasized that China, like Japan, needed to master mining, industry, and commerce. Because machines had augmented the power of European states and enhanced the welfare of the people, Kang contended that the Qing dynasty had to change its goals. He advocated educating the people in technology and not just building factories and arsenals based on foreign models. In making their case, Kang and the reformers often demeaned the results achieved when the Foreign Affairs Movement had promoted industrialization from 1865 to 1895.

In the post-1895 political environment, reformers claimed that they were championing unprecedented policies, when in fact their calls for science and industry built on the efforts of their predecessors. Where Kang Youwei and others did break new ground, however, was in their demand that traditional subsistence agriculture be mechanized. A new industrial-commercial society was the goal. In this endeavor, the reformers were as impractical as the self-strengtheners had been a generation earlier. Kang's focus on an educational transformation that would increase the numbers of those trained to industrialize China through science and technology was on target, however. By 1905, the dynasty had abrogated the old civil examinations, and most literati now faced new career expectations that drew them away from the classical curriculum that Chinese families had esteemed until then.[9]

In particular, Kang was influenced by late Qing translations of Western political economy, such as Joseph Edkins's *Policies for Enriching the Dynasty and Nourishing the People,* which was included in the *Primers for Science Studies* published in Beijing and later reprinted in Shanghai. The missionary Timothy Richard (1845–1919), who unlike Martin and Fryer had confidence in the Qing reforms under way since the Sino-Japanese War, also influenced Kang. In 1895, Richard published a series of forward-looking essays on policy matters, which the reformers published in Shanghai under the title *Tracts for the Times.* Kang Youwei and Liang Qichao consulted with Richard and drew on his essays for inspiration during the 1898 reforms.[10]

Kang was also influenced by the science translations from the Jiangnan Arsenal when he visited Shanghai in 1882. As a result, Kang established mathematics as part of the curriculum in his Guangzhou academy, where in the early 1890s he tried to apply its geometrical axioms to the political and philosophical views he had laid out in his *Complete Book of True Principles and Public Laws.* Likewise, Tan Sitong as an amateur mathematician and advocate of science established a mathematics academy in 1895, one of the first such academies. A typical reformer, Tan believed that mathematics was the foundation for both science and technology. Tan also read both traditional and Western mathematical works and became interested in geometry and algebra. In his major published work, *Studies of Benevolence,* for example, Tan relied on mathematics as an authority more than as a tool to reveal the unity of all learning.[11]

When Tan traveled to Beijing in 1893, he had his first contact with Westerners. He saw Fryer in Beijing in 1896, who introduced him to fossils, adding machines, the X-ray, and a device that purported to measure brain waves. Tan was most impressed with Fryer's translation of a work on psychology by Henry Woods (1834–1909) entitled *Ideal Suggestion through Mental Photography,* which later informed Tan's stress on mental power as the source for moral cultivation. In *Studies of Benevolence,* Tan declared the axiom: "Ether and electricity are simply means whose names are borrowed to explain mental power."[12]

Tan's views of the ether were drawn from trends in physics before 1905 when electromagnetic fields were not understood. Both William Thomson (Lord Kelvin) and James Maxwell (1831–1879), for example, thought that ether was the key to a physical theory that explained electromagnetic phenomena. In addition, modern Chinese fiction writers appropriated the translations of modern science to present a technological utopia in late Qing science fiction novels.

Tan's appropriation of the ether paralleled the attempt by European physicists to affirm ether as the fundamental unity of spiritual and material phenomena. Balfour Stewart, whose English primers on physics Young J. Allen in 1875 and Joseph Edkins in 1886 had translated into Chinese, had claimed in a coauthored 1875 volume entitled *The Unseen Universe or Speculations on a Future State* that the human personality survived as a spiritual entity after death in a parallel universe. Similarly, Maxwell and his supporters believed that ether was not reducible to any known physical substance. In China, publicists such as Tan Sitong and Kang Youwei drew on this view of ether as an active medium in the universe that resembled the traditional notion of qi. Reformist works appropriated the methods, logic, and nomenclature of science.[13]

From Traditional to Modern Mathematics

Since 1865, Li Shanlan and Hua Hengfang had presented Chinese students with an amalgam of traditional Chinese mathematics and modern mathematics. But Chinese mathematicians increasingly realized that Western mathematics had evolved independently of traditional methods. Ding Fubao attacked the claims of Chinese origins most strongly. At the same time, however, literati mathematicians tried to explain the convergence of Chinese and Western mathematics despite their separate origins. For this, they used a universalistic argument borrowed from classical learning that "the same mind produces the same principles" to explain Western developments.[14]

Other mathematicians still taught Chinese methods in mathematics until the 1901 reforms of the civil service examinations forced them to use Western mathematics for teaching problem-solving. After the reforms added the new fields of foreign arts and sciences, the questions in the mathematics section presumed knowledge of Western mathematics—so students needed training in the new field. Zou Zunxian's 1904 textbook, *Applying Algebra to Various Types of Problems,* was prepared for an accelerated course in mathematics to fill this need, but at the same time he merged Chinese and Western mathematics: his text presented Chinese solutions first so the students would be aware of traditional methods before using modern algebra to solve traditional problems.

By 1905, however, algebra and the calculus had completely replaced traditional mathematics. Since the late 1890s, the annual examinations on mathematics held at the Beijing School of Foreign Languages no longer included questions on traditional equations or procedures, although Chinese scientists still carried out research using traditional Chinese mathematics until 1900. (By contrast, Japanese mathematicians had mastered a Western-only form of mathematics after 1868.)[15]

When the Qing government promulgated its New Governance educational system by establishing primary, middle, and high schools in 1902, the reformed curriculum imitated the program initiated at the Imperial University of Beijing after 1895. There were seven fields of learning: politics, literature, science, agriculture, industry, commerce, and medicine. Science was subdivided into six areas: astronomy, geology, arithmetic, chemistry, physics, and zoology-botany. Few teachers were immediately available to teach such specialty subjects, however.[16]

In 1905, when even the civil examinations were eliminated, further regulations put into place a fresh curriculum and textbooks for the new schools. Via the reform of the education system, China after 1905 was fully converted to Western mathematics. The Qing government rearranged curricula according to the four Western school levels for mathematics courses:

1. Junior primary school: four arithmetical operations and decimals

2. Senior primary school: fractions, ratios, areas, and volumes
3. Middle school: algebra, geometry, and trigonometry
4. High school: analytical geometry and calculus

This changeover entailed the westernization of mathematics textbooks and using Western formats for numbers (1, 2, 3), known quantities *(a, b, c)*, unknown quantities *(x, y, z)*, and derivatives *(dy/dx)* and integrals ($\int$ydx). The traditional notational forms for solving equations with single or four unknowns faded into obscurity.[17]

Modern intellectuals such as Du Yaquan, Ding Wenjiang (1887–1936), and Cai Yuanpei (1868–1940) quickly left behind the typical classical education. Yan Fu, whose poor prospects in the civil examinations had spurred him to enter the Fuzhou Navy Yard's School of Navigation in 1866, associated the power of the West with modern schools where students required practical training in the sciences and technology. For Yan Fu and the post-1895 reformers, Western schools and Westernized Japanese education were examples that China should emulate. A new focus on science education for large numbers of Chinese scholars seemed to promise a way out of the quagmire of the imperial education and civil examination regime, whose educational efficiency was suspect in the 1890s. By 1911, middle and high schools were required by the Ministry of Education to evaluate students in ten areas of instruction:

1. Philosophy
2. Chinese literature
3. World literature
4. Art and music
5. History and government
6. Mathematics and astronomy
7. Physics and chemistry
8. Animal and plant biology

9. Geography and geology
10. Sports and crafts

Overall, until 1923 the new educational system mandated required courses in five areas of specialization: language, social sciences, natural sciences, mathematics, and engineering.

Among those affected by the educational changes, Ren Hongjun (1886–1961) helped found the Science Society of China in 1914 while studying at Cornell University. He had passed the last county civil examinations in 1904, and by 1907 he was a student in Shanghai where he met the future cultural iconoclast Hu Shi (1891–1962). Ren then traveled to Japan in 1908 and in 1909 entered the Higher Technical College of Tokyo as a student subsidized by the Qing government. The Qing dynasty had reached a fifteen-year agreement with the college to send forty students annually to Tokyo.[18]

While in Tokyo, Ren also joined Sun Yat-sen's early political partisans and rose to an important position in the Tokyo Sichuan branch of the emerging Nationalists. When he returned to China after the 1911 revolution, Ren served in Sun's provisional government. He received a Qinghua University fellowship in 1912 to study chemistry at Cornell. Ren completed his bachelor's degree in chemistry from Cornell, where he studied from 1912 to 1916, and his master's in chemistry from Columbia in 1917, during which time he assumed a leading role in forming a Chinese science organization that would replace the old-style literary societies.[19]

Modern Medicine in China

Those trained in modern Western medicine derided classical Chinese medicine, which was the largest field of the Chinese sciences during the 1895–1911 transition from the late Qing to the Republican era. Traditional physicians were more successful in retaining their prestige than Chinese astronomers, geomancers, and alchemists, who were dismissed by most modern scholars for practicing allegedly superstitious forms of knowledge. Chinese scholars increasingly called for a cosmopolitan synthesis of Western experimental procedures with traditional Chinese medicine.

In 1884, for instance, Tang Zonghai pointed to what he considered the dismal state of Chinese medicine in his *Convergence of the Essential Meaning of Chinese and Western Medicine to Explain the Classics* (Shanghai, 1884, 1892). In about 1890, Yu Yue (1821–1907) followed up with the first overall attack on the ancient medical practices. The book, which he entitled *On Abolishing Chinese Medicine,* may have been prompted by the troubling deaths of his talented wife and children due to illness. Using evidential research methods, Yu concluded that the oldest Chinese materia medica was valueless, and he contended that there were no essential differences between popular priests and physicians. Yet despite his support for Western medicine, Yu Yue still critiqued Western science as derivative.[20]

Chinese physicians of traditional medicine, however, remained large in number and very influential in China despite inroads made by missionary physicians and Western hospitals and the success of anatomy in mapping the internal venues for bodily illnesses. Western-style doctors and Chinese physicians, who had coexisted since 1850, had remarked on the limitations in each other's theories of illness and therapeutic practices, but for the most part each focused on its own caregiving traditions. Moreover, until anesthesiology was introduced in the early twentieth century and miracle drugs were discovered in the 1940s, the curative power of Western medicine, especially surgery, remained problematic when compared to the noninvasive pharmacopoeial traditions of Chinese physicians.[21]

One aim of the New Governance policies after 1901 was a plan—never realized—to increase Qing state involvement in policing public health. State involvement in local health issues since the late Ming had markedly diminished. Local elites filled the gap by making charitable contributions to deal with health emergencies at a time when literati also took an increased interest in medical knowledge. Late Qing public health policies signified an intention to break with this practice of abandoning medical issues to local gentry and Chinese literati-physicians.

By the end of the nineteenth century, the Qing government increasingly saw its role in statist terms. Treaty ports such as Shanghai and Tianjin became venues that linked local elite initiatives to the increasing numbers of foreigners who favored sanitation reform and public health procedures. Using Germany as a model, which Meiji Japan had also emulated, the Qing government began to use quarantine and isolation hospitals to deal with epidemics of infectious disease. (During the Ming and Qing, by contrast, when local physicians had faced southern epidemics, the state was only minimally involved.)[22]

Qing public health policies accomplished little, however, until an epidemic of plague in northeast China took some sixty thousand lives from 1910 to 1911. The Qing state turned to Wu Liangde (1879–1960) to bring the epidemic under control. Trained in medicine at Cambridge University, Wu dramatically demonstrated through substantial immunization and exacting quarantine measures the superiority of Western medicine to officials and the public. As a result, the slow emergence of the modern Chinese state during the Qing-Republican transition was tied to the extension of Western medicine and the appropriation of Western models for state-run public health systems.[23]

One byproduct of government involvement in public health was that Western-style physicians and classical Chinese doctors organized into separate medical associations, drawing the state into the contest for medical legitimacy between them. Hence, the modernizing state was progressively tied to Western medical theories and institutions, while Western-style doctors controlled the new Ministry of Public Health. When the Nationalist-sponsored Health Commission proposed to abolish classical Chinese medicine in February 1929, however, traditional Chinese doctors immediately responded by calling for a national convention in Shanghai on March 17, 1929, which was supported by a strike of pharmacies and surgeries nationwide. The protest succeeded in halting the proposed abolition, and the Nationalist government subsequently established the Institute for National Medicine to address traditionalists' concerns. One of its objectives, however, was to reform Chinese medicine along Western lines.[24]

The consequences of increased state involvement in medical policy after 1901 were significant for both Western and Chinese medicine. After 1929, the Republican government established two parallel institutions, one Western and one Chinese, each politically and educationally distinctive. This dichotomy survived both the Nationalist Republic and the Communist People's Republic, which endowed "traditional Chinese medicine"—a practice that did undergo some modernization in the early decades of the century—with an institutional, educational, and occupational base distinct from that of Western medicine.[25]

The influence of Western medicine in early Republican China presented a substantial challenge to traditional Chinese doctors. The practice of Western medicine in China was assimilated by individual Chinese doctors in a number of different ways. Some defended traditional Chinese medicine, but sought to update it with Western findings. Others tried to equate Chinese practices with Western knowledge, lumping them together as medical learning. The sinicization of Western pharmacy by Zhang Xichun (1860–1933), for example, was based on the rich tradition of pharmacopoeia in the Chinese medical tradition. Another influential group associated with the Chinese Medical Association, which stressed Western medicine, criticized traditional Chinese medical theories as erroneous because they were not scientifically based.[26]

In this cultural encounter, Chinese practitioners such as Cheng

Dan'an (1899–1957) modernized techniques such as acupuncture. Cheng's research enabled him to follow Japanese reforms by using Western anatomy to redefine the location of the needle entry points. His redefinitions of acupuncture thus revived what had become from his perspective a moribund field that was rarely practiced in China and, when used, also served as a procedure for bloodletting. Indeed, some have argued that acupuncture may have originally evolved from bloodletting.[27]

This Western reform of acupuncture, which included replacing traditional coarse needles with the filiform metal needles in use today (Figures 7.1 and 7.2), ensured that the body points for inserting needles were no longer placed near major blood vessels. Instead, Cheng Dan'an associated the points with the Western mapping of the nervous system. A new scientific acupuncture influenced by Japan and sponsored by Chinese research societies thus emerged alongside traditional acupuncture, providing with its better map of the human body an enhanced diagnosis of its vital and dynamic aspects.[28]

Similarly, Chinese doctors assimilated the discourse of nerves and the theory of germ contamination from Western medicine. These new views provided ways for Chinese physicians to discuss older illnesses such as leprosy, depletion disorder, or the wasting sickness. As a description of debilitated nerves, sexual neurasthenia now explained the illness that Chinese physicians associated with the depletion of the body's vital essences. Multiple interpretations of germ theory enabled Chinese to equate the attack of tuberculosis germs as a contingent, external cause that had been brought on by the susceptibility of a weakened body whose natural vitality had wasted away.[29]

Meiji Japan's Influence on Modern Science in China

In the late nineteenth century, Western learning exposed the Chinese to the limits of traditional categories for scientific terminology.

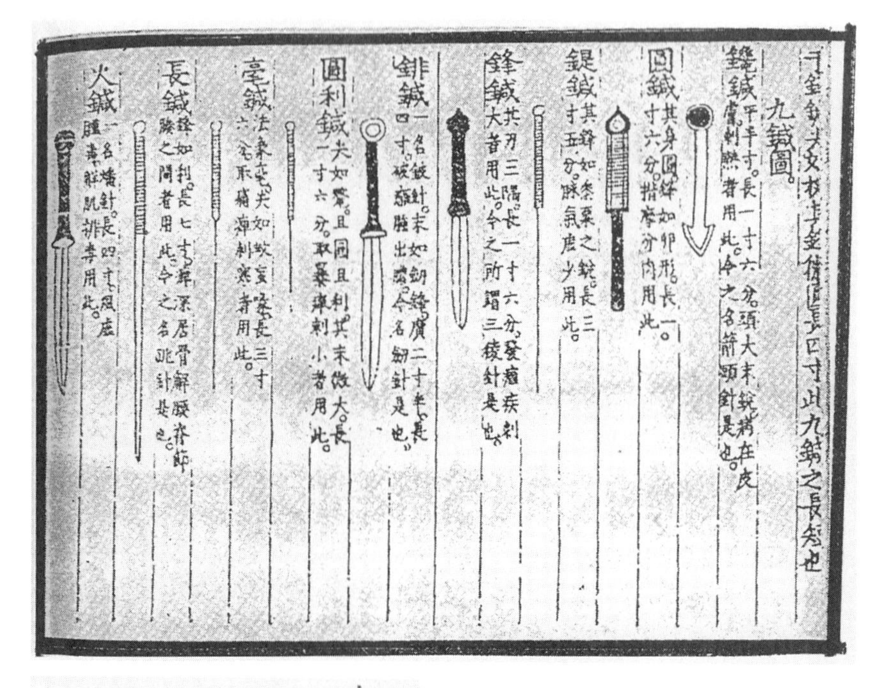

九鍼圖。

鑱鍼 平半寸。長一寸六分。頭大末銳。病在皮膚刺熱者用此。今之名箭頭針是也。

圓鍼 其身圓。鋒如卵形。長一寸六分。揩摩分肉用此。

鍉鍼 其鋒如黍粟之銳。長三寸五分。脈氣虛少用此。

鋒鍼 其刃三隅。長一寸六分。發痼疾刺大者用此。今之所謂三稜針是也。

鈹鍼 一名鈹針。末如劍鋒。廣二寸半。長四寸。破癰腫出膿。今名劍針是也。

圓利鍼 尖如氂。且圓且利。其末微大。長一寸六分。取暴痹刺小者用此。

毫鍼 法象毫。尖如蚊虻喙。長三寸六分。取痛痹刺寒者用此。

長鍼 鋒如利。長七寸。痹深居骨解腰脊節腠之間者用此。今之名跳針是也。

火鍼 一名燔針。長四寸。風虛腫毒解肌排毒用此。

Figure 7.1. Traditional coarse needles used in premodern acupuncture. *Source: Zhenjiu daquan,* 1601.

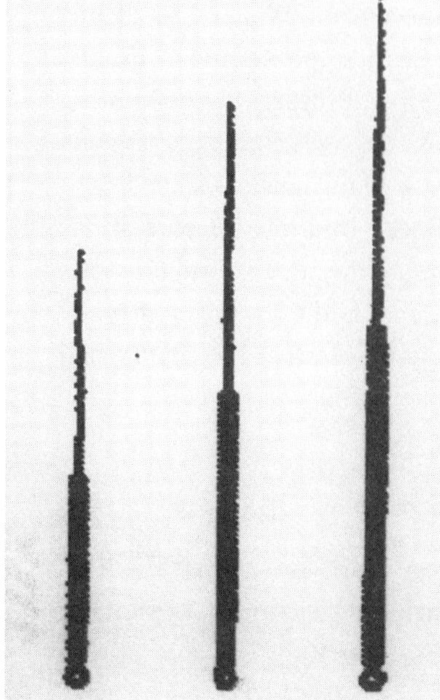

Figure 7.2. Modern filiform metal needles widely used in contemporary acupuncture clinics. They are made of gold, silver, alloy, and so on, but most are made of stainless steel.

Increasingly, literati found the claim that Western learning derived from ancient China unacceptable. Instead, many turned to Japanese terminology for the modern sciences to break with the Chinese past. By 1903, state and private schools were increasingly borrowing from Japanese translations to enunciate the modern classifications of the social sciences, natural sciences, and applied sciences.[30]

The Effect of Qing Science Translations on Japan

Before 1894, Japan had imported many European books on science from Qing China, particularly after 1720 when the shogun Yoshimune (r. 1716–1745) relaxed the Tokugawa prohibition of all Christian books. Many were translated into Chinese after the Japanese expelled the Jesuits for meddling in their late-sixteenth-century civil wars. Ricci's *mappa mundi,* Chinese translations of Euclid's geometry, and Tychonic astronomy, for example, all made their way to Tokugawa Japan.[31]

Works from late Ming collections and Kangxi-era compendia arrived in Japan via Ningbo-Nagasaki trade and from Fujian after the 1720s. The Japanese also avidly adopted eighteenth-century Chinese terminology for Sino-Western mathematics when the second edition of Mei Wending's complete works was imported in 1726 and translated two years later. In addition, physics, chemistry, and botany books, imported from Europe via the Dutch trading enclave in Nagasaki harbor in the early nineteenth century, were translated from Dutch into Japanese.[32]

The Meiji government also was drawn to translations on science prepared under the auspices of Protestant missionaries in the Chinese treaty ports. Prominent translations into Chinese of works dealing with symbolic algebra, calculus, Newtonian mechanics, and modern astronomy quickly led to Japanese editions and Japanese translations of these works. Macgowen's 1851 *Philosophical Almanac* and Hobson's 1855 *Treatise of Natural Philosophy,* for example, came out in Japan in the late 1850s and early 1860s. And four of Hobson's 1851–1858 medical works came out in Japan between 1858 and 1864.

Issues from the 1850s *Shanghae Serial* were also republished in Japan, along with translations completed in the Jiangnan Arsenal and publications from the Beijing School of Foreign Languages. The Wylie-Li translations of algebra (completed in 1859 in China, 1872 in Japan), calculus (1859, 1872), and Martin's *Natural Philosophy* (1867, 1869) were all quickly available to scholars and officials in Meiji Japan. Arguably, these works had greater influence in Japan than in China.

Many Japanese scholars in early Meiji times still preferred Chinese scientific terms over translations derived from Dutch Learning. The Chinese word for chemistry *(huaxue)*, for example, replaced the term *chemie* (*semi* in Japanese) derived from the Dutch. Similarly, the influence of Jiangnan Arsenal publications can be seen in the choice of Chinese terminology for metallurgy *(jinshi xue)* used in Japanese publications, a phrase that was later changed in Japan and reintroduced to China as a new term for mining *(kuangwu xue)*.

Japan's famous Iwakura mission to tour the world stopped in Shanghai in September 1873 at the end of its journey to Europe and the United States. The members took a tour of the Jiangnan Arsenal and described the shipyard, foundry, school, and translation bureau in very positive terms. The mission noted how the shipyard was operated initially by British managers, with the aid of Chinese workers who had trained abroad. The account added that "now the entire management of the yard is in the hands of Chinese" and concluded: "This one yard would be capable of carrying out any kind of work, from ship repair to ship construction."[33]

When the diplomat Yanagihara Sakimitsu (1850–1894) visited China, he purchased many of the Chinese scientific translations. On his third visit in 1872, for instance, he bought twelve titles on science and technology in thirty-one volumes from the Jiangnan Arsenal. These included works on chemistry, ship technology, geography, traditional mathematics, mining, and Chinese trigonometry. The Japanese government continued to buy arsenal books until 1877. In 1874, Yanagihara received twenty-one newly translated books from China. Yet despite the influence of Dutch Learning and translations from China, and even though the Japanese began teaching modern Western science on a large scale in the 1870s, the Chinese did not borrow many scientific terms from Japan before the Sino-Japanese War.

Unlike the Chinese translations that were readily transmitted to Japan, Tokugawa authorities kept Dutch Learning translations secret. While much has been made of the contributions of Dutch Learning to Japanese science during the Tokugawa period, there is no evidence that Dutch Learning per se enhanced or determined the course of Meiji science and technology. Moreover, the influence of Dutch Learning, while important among samurai elites in the late eighteenth and early nineteenth centuries, was not sufficient to touch off in Tokugawa Japan the sort of technological revolution based on Newtonian mechanics and French analytical mathematics that was described in Chapter 3 as the engineer's toolkit.

Japan's overwhelming triumph in the Sino-Japanese War suggested a scientific advantage over the Qing dynasty that most accounts since 1895 have simply assumed derived from the influence of Dutch Learning. The Yokosuka Dockyard, for instance, was still dependent on French engineering advisers until the 1880s and on British technical aides in the 1890s. In fact, when Japanese leaders founded the Yokosuka Dockyard in 1867 to replace late Tokugawa regional dockyards in Nagasaki and elsewhere, they used the design of French architect-engineer François-Léonce Verny (1837–1908), who had graduated from the Ecole polytechnique in 1858. Verny worked on shipbuilding projects in Ningbo and Shanghai in China for two years before coming to Japan in 1865. In the midst of the

Meiji Restoration in 1868, his Japanese superiors at Yokosuka thought he should choose a Yokosuka architect to take a short leave to study the Fuzhou Navy Yard, which the Qing government was then building with French technical advice. In 1871, the construction director at Fuzhou visited Verny in Yokosuka, now under the Meiji Public Works Ministry. Verny noted that the Fuzhou Navy Yard's budget was three times his.[34]

Japanese Science in China after 1895

Beginning in 1896, Chinese scholars translated science books that the Japanese had prepared themselves. By 1905, the new Qing Ministry of Education was staunchly in favor of science education using textbooks from the Japanese scientific system. Instead of receiving and translating Western works directly, Chinese literati and officials would now use Japanese works to mediate Western developments for them.[35] We have seen that after the Sino-Japanese War, reformers encouraged Chinese students to study in Japan. Kang Youwei in particular enthusiastically promoted Meiji Japan scholarship in his *Annotated Bibliography of Japanese Books* and in his reform memorials to the Guangxu emperor (r. 1875–1908). Kang recommended 339 works in medicine and 380 works in the sciences, which now replaced lists of the best Western books. The Guangxu emperor's edict of 1898, too, encouraged study in Japan.[36]

As a publicist in exile in Japan, Liang Qichao translated Japanese materials into Chinese at a fast clip. Luo Zhenyu (1866–1940) was also a prolific translator. In addition to his antiquarian interests, he published the *Agricultural Journal* from 1897 to 1906 in 315 issues. The articles were mainly drawn from Japanese sources on science and technology. Luo also compiled the *Collectanea of Agricultural Studies* in eighty-eight works, with forty-eight based on Japanese books. Du Yaquan, too, edited journals in 1900 and 1901 that translated science materials from Japanese journals—and that had the distinction of being the first science journals edited solely by a Chinese scholar. Meanwhile, the massive translation in Shanghai of a Japanese encyclopedia took several years. When it appeared in 1904, the encyclopedia contained over one hundred works, with twenty-eight in the sciences and nineteen in applied science.

Post-Boxer educational reforms of 1902–1904 were also crucial for the transformation of education in favor of Japanese-style science and technology. The civil examinations were the last bastion of traditional Chinese science, where the "Chinese origins" approach to Western learning remained obligatory. After the examination system was abolished in 1905, Japanese science texts became the new models for Chinese education at all levels of schooling. From 1886 to 1901, for instance, the Meiji government officially approved eleven different texts on physics, and the Chinese translated eight of these for native editions. From 1902 to 1911, twenty-two different physics texts were approved in Japan, and the Chinese translated seven.

Similarly, in chemistry from 1902 to 1911, the Chinese translated seventy-one Japanese texts, mostly for use in middle schools and teacher's colleges. Japan produced twelve middle-school chemistry texts between 1886 and 1901. Of these, the Chinese translated six. And of the eighteen middle school chemistry texts produced by the Japanese between 1902 and 1911, five were translated into Chinese. Japanese scientists were also invited to lecture in China, and Chinese translators, with the help of Japanese educators, worked on Iimori Teizō's (1851–1916) edited volume *Physics* for the Jiangnan Arsenal from 1900 to 1903. Iimori's wide influence on Chinese physics grew out of this project. Further, by 1907, when Yan Fu had taken charge of the Qing Ministry of Education's committee for science textbooks, he had approved the use of Japanese scientific terms.

We should not underrate the historical importance of Japanese translations for the development of modern science in China. Japanese translations were much more widely available in China than those produced earlier. In addition, the new Japanese science textbooks contained newer content than the 1880s Jiangnan Arsenal and missionary translations, which were already outdated by European standards in the 1890s. The introduction of post-1900 science via Japan, which included new developments in chemistry and physics, went well beyond what Europeans and Americans had provided to the emerging Chinese scientific community.[37]

Chinese presses also published translations of Japanese texts in greater numbers, and they were easier to read because only Chinese compiled them. Moreover, the quality of translations from works by Japanese scientists was better than those of the earlier science primers, because Chinese translators themselves could understand the Japanese originals. In addition, the Japanese texts were available to a wider audience of students in the new public schools and teachers' colleges that the Qing government had established after 1905. Further facilitating the flow of information from Japanese to Chinese, the Imperial University in Beijing invited Japanese professors to join its faculty.[38]

Finally, to make the new translations more easily understood than standard classical translations, Chinese translators helped produce a new literary form for presentation of the sciences, which contributed to the rise of the vernacular for modern Chinese scholarly and public discourse. Among the urbanite generation that matured during the New Culture Movement of 1915–1919 and the May Fourth era after 1919, especially in Beijing and Shanghai, the first decade of the twentieth century provided a basic education in modern science via Japanese textbooks.

The Delayed Emergence of Physics as a Technical Field

When we compare the development of modern physics in Meiji Japan and late Qing China, we find that scholars in both countries had started to master Western studies in the early and mid-nineteenth century. The Translation Bureau at the Jiangnan Arsenal and the Dutch Translation Bureau in Tokugawa Japan produced books on physics beginning in the 1850s in China and from 1811 in Japan. Although the introduction of Dutch Learning in the seventeenth and eighteenth centuries enabled an earlier start in Japan, the materials on physics in the Protestant translations produced in China after 1850—quickly transmitted to Japan—made those earlier studies out-of-date. Moreover, the primer series produced in the 1870s and early 1880s in China remained superior overall to their Meiji counterparts until the 1890s.

Despite the range of science translations in Qing China through the 1880s, physics textbooks were not available in China until they were first published in Japan, in large part because of the way Protestant missionaries had introduced the physical sciences in China. Missionary translators first introduced the disaggregated branches of physics rather than a unified field of physics. Accordingly, mechanics, optics, acoustics, electricity, and thermodynamics were presented independently. By presenting only the subfields of physics, the translators made it difficult for Chinese to appreciate the unity of the field. Moreover, introducing the branches first made it more complicated later to reach a consensus for physics.

Often physics was equated with investigating things. Others preferred calling physics investigating things and extending knowledge, which frequently overlapped vaguely with the general term for science and created substantial misunderstanding. Edkins's 1886 *Science Primers,* for example, associated "investigating the materiality of things" with physics. In 1895, the physics program in the Beijing Foreign Language School changed its name from the Hall for Investigating Things to the Hall for Investigating and Extending Knowledge.[39]

The Qing state was also slower in reforming its educational system. Meiji Japan's new educational system was established in 1868.

The Qing initiated comparable educational reforms in 1902. The Japanese formed their Ministry of Education in 1871, while the Qing established its counterpart in 1905. Similarly, Japan founded Tokyo University as the key modern teaching institution in 1877, but the Imperial University of Beijing did not exist until 1898. Courses in physics had already started in 1875 in Japan when the Tokyo school that evolved into the university shifted from foreign language lectures to lectures in Japanese by returned physics students. The first Japanese students trained in Japan graduated in physics in 1883.

Chinese science faculties were not established at the Imperial University of Beijing until 1910, but even then only classes in chemistry and geology were taught. Physics was added in 1912. Beijing recruited Japanese science teachers to the university starting in 1902, but they left in 1908–1909 after their six-year contracts expired. From 1898 to 1911, only two hundred students were trained in the sciences at the Imperial University, and the initial absence of faculties of mathematics and physics remained a serious problem for training scientists. In Japan, there were few students of physics when compared to the more popular fields of law and medicine, but between 1882 and 1912 Tokyo University still graduated 186 in physics.

Japan's educational system had a head start in editing and translating physics textbooks. China by comparison lacked textbook materials to teach physics at all levels of the education system. Similar delays occurred in other technical fields such as chemistry and geology. By 1873, the Japanese were teaching physics in the new Meiji schools, and Tokyo University had a physics program from 1877. By contrast, the Beijing School of Foreign Languages asked only occasional physics questions on examinations from 1868, which were based on Martin's elementary textbook. The subfields of physics, which were taught in some military and arsenal schools as well, were taught separately as mechanics, hydraulics, acoustics, pneumatics, heat, optics, and electricity.

Meiji educators produced physics textbooks in the 1870s, but none were available in China until the 1890s. Although the Japanese relied initially on Protestant translations from China, the Education Ministry ordered Katayama Junkichi (1837–1887) to compile an official physics textbook when physics became a specialized discipline. Katayama's textbook was added to the Japanese curriculum in 1876 and republished many times. Moreover, Japan invited Western scientists to Japan. The influence of Western publications was also significant: Iimori Teizō completed his edition of *Physics,* for example, by consulting physics works published in Germany.

As a result of the 1898 reforms, the Qing government decided to copy the Meiji model for education and create a public school system for science education, rather than simply rely on schooling in the arsenals, navy yards, and factories. Full implementation of this program was not feasible until 1904–1905, when the civil examination system was scrapped in favor of the new school system. The Sino-Japanese War had taught the Qing government that it was not enough to rely solely on the arsenals for this important education.[40]

Direct Chinese translations of the best physics texts by the most famous Japanese physicists, such as Iimori's *Physics,* became the most efficient means in the early twentieth century to prepare textbooks for the new Qing schools. This policy also guaranteed that the Chinese would no longer rely on Western missionary informants for translations in important, specialized fields such as physics. But China's dependency on Japan was reconsidered after 1915 when Japan's policies toward the Republic of China became increasingly predatory.

Although high-level education in physics began at the Beijing Imperial University in 1912, the best-trained physicists studied in the United States and Japan. When Beijing University was reorganized in 1912, it had formal divisions between the humanities and the sciences, which included the three fields of mathematics, chemistry, and physics. An independent physics department was not created until 1917, but the greater scope of physics texts in the school system after 1905 did provide for wider knowledge of the field in China than before 1900.

By 1900, Japan also had a lead over China in physics research, the unification of technical terminology, and the number of research associations. For instance, Japanese scholars started publishing in physics in 1880s, leading to the publication of over two hundred articles in the various subfields of physics by the end of Meiji era in 1912. Moreover, several Japanese physicists emerged who approached Western levels of expertise.

Meiji translators first unified the terms for physics in the 1870s after they chose the official designation for physics in 1872. Terminology in Japanese physics achieved a final unification with the 1888 publication of an official list of technical terms with foreign counterparts. The committee for systematizing the translation of terms for physics, which began in 1885, was led by three of the first Japanese graduates in physics from Tokyo University. Scholars unified terms for a total of 1,700 items from English, French, and German, which they then translated into Japanese and published. The Chinese started using the Japanese term for physics in 1900 when Iimori's book by that name was published in China. Before then, the equivalent Chinese term had usually referred to the traditional fields of natural studies.[41]

Academics created the first mathematics society in Tokyo in 1877 with fifty-five members. In 1884, ten of its seventy-five members specialized in physics. When Japanese formed the Tokyo Mathematics-Physics Society in 1884, it started with eighty-two members, twenty-five of whom were physicists. The group changed its name in 1919 to the Japan Mathematics-Physics Society, which survived as an organization until it separated into two parts in 1948. Smaller specialized groups in physics were also formed in Japan in the 1880s.

China was also behind Japan in training physicists and organizing associations. Chinese scholars had to study physics abroad, and the research institutes for physics at the Academia Sinica, the Beijing Institute, and the Qinghua Institute were not formed until 1928–1929. Although Chinese terms for physics were unified in 1905, they were not finally settled until the 1920s. Moreover, the Chinese Science Society and its journal were not founded until 1915, and that took place abroad, at Cornell University in the United States. Physicists did not form the Chinese Physics Society until 1932.

The belief that Western science represented a universal application of objective methods and knowledge was increasingly articulated in the journals associated with the New Culture Movement after 1915. The journal *Science*, which the newly founded Chinese Science Society created in 1914, assumed that an educational system based on modern science was the panacea for all of China's ills. Meiji Japan served as the model for that cure-all until 1915 when Japanese imperialism, like its European predecessor, forced Chinese officials, warlords, and intellectuals to reconsider the benefits of relying on Japan.

Forgetting China's Early History of Science

Despite the late Qing curriculum changes described earlier, which had prioritized science and engineering in the new public schools since 1902 and in private universities such as Qinghua, many Chinese university and overseas students were by 1910 increasingly radical in their political and cultural views, which carried over to their convictions about science. Traditional natural studies became part of the failed history of traditional China to become modern, and this view now asserted that the Chinese had never produced any science. How the premodern Chinese had demarcated the natural and the anomalous vanished when modernists, anarchists, republicans, and socialists in China accepted the West as the universal starting place of all science.[42]

After 1911, many radicals linked the necessity for a Chinese political revolution to the claim that a scientific revolution was also mandatory. Those Chinese who thought a revolution in knowledge required Western learning not only challenged classical learning, but also unstitched patterns of traditional Chinese natural studies and medicine long accepted as components of imperial orthodoxy.

As Chinese elites turned to Western studies and modern science, fewer remained to continue the traditions of classical learning or moral philosophy that had been the basis for the imperial orthodoxy and the civil examinations before 1900. Those who still focused on traditional learning, such as the pioneering historian Gu Jiegang (1893–1980) in Beijing and others elsewhere, often did so by reconceptualizing ancient learning in light of "doubting antiquity" and applying new, objective procedures for historiography that they had derived from the sciences. Thereafter, the traditional Chinese sciences and classical studies survived only as vestiges of native learning in public schools established by the Ministry of Education after 1905. They have endured as contested scholarly fields in universities since 1911.[43]

The Great War from 1914 to 1919 profoundly challenged both those modernists in China who thought science a universal model for the future and the "New Confucian" traditionalists who revived Chinese moral teachings after the devastation visited on Europe. The reformer cum scholar-publicist Liang Qichao, who was then in Europe leading an unofficial group of Chinese observers at the 1919 Paris Peace Conference, visited a number of European capitals. His group witnessed the effect on Europe of the war's deadly technology. They also met with leading European intellectuals, such as the German philosopher Rudolf Christoph Eucken (1846–1926) and the French philosopher Henri Bergson (1859–1941), to discuss the moral lessons of the war.[44]

In his influential *Condensed Record of Travel Impressions While in Europe,* Liang Qichao related how the Europeans whom they met regarded World War I as proof of the bankruptcy of the West and the end of the "dream of the omnipotence of modern science." Europeans now sympathized with what they considered the more spiritual and peaceful "Eastern civilization" and bemoaned the legacy in Europe of an untrammeled material and scientific social order that had fueled the world war. Liang's account of the spiritual decadence in postwar Europe indicted the materialism and the mechanistic assumptions underlying modern science and technology. A turning point had been reached, and the dark side of what New Culture enthusiasts called "Mr. Science" had been exposed. Behind it lay the colossal ruins produced by Western materialism.[45]

The largest archive of premodern records for the study of nature remains in China. By better understanding the history of imperial Chinese natural studies, technology, and medicine, and the cultural system that undergirded them, we can form a more nuanced perception of the belief systems that inform our contemporary versions of modern science. It is essential to probe the surface of self-satisfied rhetoric about science as fundamentally Western and to go beyond simplistic appeals to the Greek deductive logic as the mother of all science.[46]

If premodern science in China, Europe, India, and the Islamic world denoted a rational and abstract understanding of the natural world, the rise of modern science in the eighteenth and nineteenth centuries melded the exact sciences with machine-driven technologies that surpassed the rich artisanal traditions of that early modern world. Frequently, historians of modern science stress this transition in light of the application of mathematical hypotheses to nature, the use of the experimental method, the geometricization of space, and the mechanical model of reality to explain the scientific revolution in the West. I am agnostic about these efforts to explain why China or the Islamic world failed to develop the rigorous mental mindset that accompanied the introduction of modern science.[47] But it is certain that with the exception of a modernized version of "traditional Chinese medicine" that flourishes globally as one version of holistic medicine, the traditional fields of natural studies in imperial China did not survive the modern science revolution between 1850 and 1920. Instead, Chinese replaced their traditional fields with the modern sciences.

The Chinese construction of modern science, medicine, and technology is a remarkable achievement, even if the Chinese did not initiate the internal and external revolutions that provoked it. Early modern Europe, after all, borrowed much from Asia and the Islamic world before its own scientific revolutions. The shock of Western and Japanese imperialism in China remains a tragedy for the Chinese, but because of the blending of natural learning that occurred from 1600 to 1900, the triumphs of the West and Japan in contemporary science, medicine, and technology should also be considered in light of contemporaneous developments in traditional and modern Chinese science.

APPENDIXES

1. Tang Mathematical Classics

2. "Science Outline Series," 1882–1898

3. Table of Contents for the 1886 *Primers for Science Studies*

4. Twenty-three Fields of the Sciences in the 1886 *Primers for Science Studies*

5. Some Officially Selected Chinese Prize Essay Topics from the Shanghai Polytechnic

6. Some Translations of Chemistry, 1855–1873

7. Partial Chronological List of Arsenals, etc., in China, 1861–1892

NOTES

ACKNOWLEDGMENTS

INDEX

APPENDIX 1: TANG MATHEMATICAL CLASSICS

1. *The Gnomon of the Zhou Dynasty and Classic of Computations (Zhoubi suanjing)* was compiled in the first century B.C.E. and reworked by commentators in medieval times. An important work for mathematical astronomy based on "enveloping heavens" *(gaitian)* cosmology, it also corroborates Pythagoras's right triangle theorem. A third-century C.E. commentary added fifteen algorithms for solving right triangles. Recovered by Dai Zhen from the *Great Compendium of the Yongle Era.*[1]
2. *Computational Methods in Nine Chapters (Jiuzhang suanshu)* was compiled from 200 B.C.E. to 300 C.E., with commentaries added later. Its 246 problems set the model for the mathematical language of linear equations *(fangcheng)* and presented the pattern for computations; it was intended as a means to apply right triangle methods *(gougu)* to calculate areas of various shapes. Recovered by Dai Zhen from the *Great Compendium of the Yongle Era.*[2]
3. *Sea Island Computational Canon (Haidao suanjing),* by Liu Hui, contains computational prescriptions for surveying inaccessible points using circle divisions with inscribed figures; it also applies the principle of proportion to right triangles. Recovered by Dai Zhen from the *Great Compendium of the Yongle Era.*[3]

4. *Sunzi's Computational Canon (Sunzi suanjing),* circa the fifth century, gives details for arithmetic operations using counting rods. Recovered by Dai Zhen from the *Great Compendium of the Yongle Era* (although a Southern Song edition also survived).[4]

5. *Computational Canon of the Five Administrative Departments (Wucao suanjing),* compiled circa the fifth century, was intended as a textbook on applied mathematics for surveying land, managing troops, collecting taxes, managing granaries, and handling money. It was added to the mathematical classics during the Song period. Recovered by Dai Zhen from the *Great Compendium of the Yongle Era.*[5]

6. *Xiahou Yang's Computational Canon (Xiahou Yang suanjing),* n.d., perhaps circa the fourth to eighth centuries, is an apocryphal collection of elementary tax problems solved through simplified computational techniques using counting rods. It was added to the mathematical classics during the Tang or Song periods. Recovered by Dai Zhen from the *Great Compendium of the Yongle Era.*[6]

7. *Zhang Qiujian's Computational Canon (Zhang Qiujian suanjing),* compiled around the period of the Northern Wei, 466–485, gives solutions for quadratic equations, arithmetic progressions, and an indeterminate problem known as the "hundred fowl problem." Recovered by Dai Zhen, based on the late Ming edition from the Southern Song dynasty.[7]

8. *Computational Rules of the Five Classics (Wujing suanshu),* compiled circa 566, was glossed by Zhen Luan circa 570 to interpret calendrical data in the Classics, and to explain the "great expansion" *(dayan)* method for solving simultaneous congruencies; it presents mythical geography, weights and measures, and musical pitch pipes. Recovered by Dai Zhen from the *Great Compendium of the Yongle Era.*[8]

9. *Computational Canon of the Continuation of Ancient Techniques (Jigu suanjing)* was compiled by calendricist Wang Xiaotang (ca. 650–750) in the seventh century. The work gives astronomical ex-

planations for construction of an astronomical tower and embankments associated with long canals where thousands of workers were mobilized. It also provides the solution of right triangles, leading to polynomial equations of the second or third degree. Recovered by Dai Zhen, based on the late Ming edition from the Southern Song dynasty.[9]

10. *Techniques for Calculations by Combination (Zhuishu)* by Zu Chongzhi (429–500). This work was annotated by Minggatu during the Qing period.

APPENDIX 2: "SCIENCE OUTLINE SERIES," 1882–1898

1. *Shengxue xuzhi* (Acoustics, first part—1)
2. *Tianwen xuzhi* (Astronomy, first part—2)
3. *Huaxue xuzhi* (Chemistry, first part—3)
4. *Dianxue xuezhi* (Electricity and magnetism, first part—4)
5. *Dili xuzhi* (Physical geography, first part—5)
6. *Dizhi xuzhi* (Political geography, first part—6)
7. *Dixue xuzhi* (Geology, first part—7)
8. *Qixue xuezhi* (Pneumatics, first part—8)
9. *Daishu xuzhi* (Algebra, second part—1)
10. *Suanfa xuzhi* (Arithmetic, second part—2)
11. *Weiji xuzhi* (Calculus, second part—3)
12. *Quxian xuzhi* (Conic sections, second part—4)
13. *Huaqi xuzhi* (Drawing instruments, second part—5)
14. *Zhongxue xuzhi* (Mechanics, second part—6)
15. *Liangfa xuzhi* (Mensuration, second part—7)
16. *Sanjiao xuzhi* (Trigonometry, second part—8)
17. *Guangxue xuzhi* (Optics, third part—1)
18. *Lixue xuzhi* (Dynamics, third part—2)
19. *Shuixue xuzhi* (Hydraulics, third part)

20. *Kuangxue xuzhi* (Mineralogy, third part—5)
21. *Quanti xuzhi* (Physiology and Anatomy, third part—7)
22. *Fuguo xuzhi* (Political economy, fifth part)
23. *Xili xuzhi* (Western etiquette—What to do, fifth part—7)
24. *Jieli xuzhi* (Western etiquette—What to avoid, fifth part—8)

APPENDIX 3: TABLE OF CONTENTS FOR THE 1886 *PRIMERS FOR SCIENCE STUDIES*

1. Western learning *(Xixue lueshu)*, by Joseph Edkins
2. Introductory for the Science Primers *(Gezhi zongxue)*, by Thomas Huxley
3. Political geography *(Dizhi)*
4. Physical geography *(Dili zhixue)*[1]
5. Geology *(Dixue)*, by Archibald Geikie
6. Botany *(Zhiwuxue)*, by J. D. Hooker
7. Physiology *(Shenli)*, by M. Foster
8. Zoology *(Dongwuxue)*
9. Chemistry *(Huaxue)*, by Henry Roscoe
10. Physics *(Gezhi zhixue)*, by Balfour Stewart
11. Astronomy *(Tianwen)*, by J. N. Lockyer
12. Political economy (*Fuguo yangmin ce,* literally, "Policies for enriching the dynasty and nourishing the people"), by W. Stanley Jevons
13. Greek history *(Xila zhilue)*
14. Roman history *(Luoma zhilue)*
15. Logic *(Bianxue)*, by W. Stanley Jevons
16. European history *(Ouzhou shilue)*

APPENDIX 4: TWENTY-THREE FIELDS OF THE SCIENCES IN THE 1886 *PRIMERS FOR SCIENCE STUDIES*

1. Astronomy
2. Physics
3. Geology
4. Zoology
5. Mineralogy
6. Electricity and magnetism
7. Chemistry
8. Climatology
9. Light
10. Mechanics
11. Fluid mechanics
12. Gaseous mechanics
13. Anatomy
14. Comparative anatomy
15. Physiology
16. Botany
17. Medicine
18. Geometry

19. Mathematics
20. Algebra
21. Calendrics
22. Archaeology
23. Folklore

APPENDIX 5: SOME OFFICIALLY SELECTED CHINESE PRIZE ESSAY TOPICS FROM THE SHANGHAI POLYTECHNIC

Spring 1887, "Theme" by Xu Xingtai, Zhejiang Provincial Administration Commissioner of Hangzhou: "Compare the sciences of China and the West, showing their points of difference and similarity."

Spring 1889, "Theme" by Gong Zhaoyuan, Zhejiang Provincial Surveillance Commissioner of Hangzhou: "What are key points in the detailed strengths and cursory weaknesses in contemporary translations of Western science?"

Spring 1889, "Extra Theme" by Li Hongzhang: "With respect to the 'Science' referred to in the 'Great Learning,' from Ching-kang-ching to the present, there have been several tens of [Chinese] scholars who have written on the subject. Do any of them happen to agree with Western scientists? Western science began with Aristotle in Greece; then came Bacon in England who changed the previous system and made it more complete. In later years, Darwin's and Spencer's writings have made it still more comprehensible. Give a full sketch of the history and bearings of this whole subject."[1]

Spring 1891, "Extra Theme" by Li Hongzhang: "Compare the similarities between the *Zhoubi suanjing* (The Gnomon of the Zhou dynasty and classic computations) and Western techniques of trigonometry for measuring segments of a circle."

Summer 1893, "Theme" by Wu Yinsun, Zhejiang Provincial Circuit Attendant for Ningbo and Shaoxing: "Try to prove in detail the following: When did Western medical techniques begin and who were they transmitted by? Are there differences in the various ways each country treats illnesses? What are the strengths and weaknesses of Chinese versus Western medical principles?"

Spring 1894, "Theme" by Nie Jigui, Shanghai Circuit Attendant and Zhejiang Provincial Surveillance Commissioner of Hangzhou: "Itemize and demonstrate using scholia that the "Jingshang" [Classic, first part] and "Shuoshang" [Expositions, first part] chapters from the *Mozi* had already raised the Western principles of calendrical studies, optics, and mechanics."

APPENDIX 6: SOME TRANSLATIONS OF CHEMISTRY, 1855–1873

Benjamin Hobson, trans., *Bowu xinbian* (Treatise of natural philosophy). Shanghai: Mohai shuguan, 1855.

William Martin, trans., *Gewu rumen* (Elements of natural philosophy and chemistry; literally, "Introduction to the investigation of things"). Beijing: Tongwen'guan, 1868.

John Fryer and Xu Shou, trans., *Huaxue jianyuan* (Mirror of the origins of chemistry). Shanghai: Jiangnan zhizaoju, 1871. The translation was completed in 1869, a sequel in 1875, and a supplement in 1882.

John Kerr and He Liaoran, trans., *Huaxue chujie* (Preliminary steps in chemistry). Guangzhou: Boji yiju, 1870.

Anatole Billequin, *Huaxue zhinan* (Guide to chemistry). Beijing: Tongwen'guan, 1873.

APPENDIX 7: PARTIAL CHRONOLOGICAL LIST OF ARSENALS, ETC., IN CHINA, 1861–1892

1. Anqing Arsenal (1861), established by Zeng Guofan
2. Jiangnan Arsenal (1865), established in Shanghai by Zeng and Li Hongzhang
3. Jinling Arsenal (1865) in Nanjing used for making breech rifles and steel; gunpowder factory from 1881
4. Fuzhou Shipyard (1866), the base for the Southern Fleet, established by Zuo Zongtang; machine factory from 1870
5. Tianjin Arsenal (1867), under Li Hongzhang used to manufacture gunpowder and acid
6. Xi'an Arsenal (1869), used to manufacture bullets and gunpowder
7. Lanzhou Arsenal (1871–1872), used to manufacture bullets and gunpowder
8. Guangzhou Arsenal (1874), used to manufacture bullets and gunpowder and to repair ships; gunpowder from 1875
9. Hunan Arsenal (1875), used to manufacture acid, gunpowder, and cannon
10. Shandong Arsenal (1875), used as a gun depot and for making acid and gunpowder

11. Sichuan Arsenal (1877), used to manufacture acid, cannon, bullets, and gunpowder
12. Jilin Arsenal (1881), used to manufacture bullets, gunpowder, and acid
13. Lüshun, Port Arthur Naval Station (1881–1882)
14. Weihaiwei Shipyard (1882) for the Beiyang Fleet
15. Beijing Field Force Arsenal (1883)
16. Shanxi Machine Shop (1884), used for making foreign gunpowder
17. Yunnan Arsenal (1884), manufacturing site for bullets and gunpowder
18. Hangzhou Arsenal (1885), used as a gun depot and for making bullets, ship mines, and gunpowder
19. Taiwan Machine Shop (1885), used for making bullets and gunpowder
20. Taiwan Arsenal (1885)
21. Daye Iron Mine (1890), in Hubei
22. Hanyang Ironworks (1890), in Hubei, established by Zhang Zhidong and used for production of iron and steel.
23. Hanyang Arsenal (1892), weapons factory
24. Zhejiang Machine Shop (1893)

NOTES

Works frequently cited have been identified by the following abbreviations:

CRZ — *Chouren zhuan* (Biographies of mathematical astronomers), comp. Ruan Yuan (1764–1849). Taipei: Shijie shuju, 1962.

DMB — *Dictionary of Ming Biography,* ed. L. C. Goodrich, et al. 2 vols. New York: Columbia University Press, 1976.

DSB — *Dictionary of Scientific Biography,* ed. Charles Gillispie et al. New York: Scribner's, 1970–1978.

ECCP — *Eminent Chinese of the Ch'ing Period,* ed. Arthur Hummel. Reprint. Taipei: Chengwen Bookstore, 1972.

GRC — *Gregorian Reform of the Calendar: Proceedings of the Vatican Conference to Commemorate Its 400th Anniversary, 1582–1982,* ed. G. V. Coyne, S.J., et al. Vatican City: Specola Vaticana, 1982.

GZHB — *Gezhi huibian* (*Chinese Scientific Magazine;* also called *Chinese Scientific and Industrial Magazine*). Reprint in 6 vols. Nanjing: Guji shudian, 1992. The dating of articles in this reprint are defective, and some sections are disordered. Reference to the original editions, though limited in number, is recommended.

GZSYKY — *Gezhi shuyuan keyi* (China Prize Essay Contest). Annual volumes. Shanghai: Shanghai Polytechnic, 1886–1893. From the Fryer Private Library, University of California, Berkeley, in 2 volumes, which include eight traditional string-bound volumes (*ce*).

GZQM — *Gezhi qimeng* (Primers for science education), trans. Young J. Allen et al. Shanghai: Jiangnan Arsenal Publications, 1879–1880.
Gezhi qimeng (Primers for science education), trans. Joseph Edkins. Beijing: Imperial Customs Office Publication, 1886.

HJAS *Harvard Journal of Asiatic Studies*
JAS *Journal of Asian Studies*
LHCT *Liuhe congtan* (*Shanghae Serial*). Published January 1857 to June 1858 by Mohai shuguan (Inkstone Press of the London Missionary Society).
SCC *Science and Civilisation in China*, ed. Joseph Needham et al. Cambridge, Eng.: Cambridge University Press, 1954–.

Introduction

1. Alain Peyrefitte's *The Immobile Empire: The First Great Collision of the East and West*, trans. Jon Rothschild (New York: Knopf, 1992), paints a dark picture of Qing cultural xenophobia and conservatism. For correctives, see Joanna Waley-Cohen, *The Sextants of Beijing: Global Currents in Chinese History* (New York: W. W. Norton, 1999), pp. 92–128.
2. See D. E. Mungello, ed., *The Chinese Rites Controversy: Its History and Meaning* (Nettetal, Ger.: Steyler Verlag, 1994).
3. Compare Catherine Jami, "Western Mathematics in China, Seventeenth Century and Nineteenth Century," in Patrick Petitjean, Catherine Jami, and Anne Marie Moulin, eds., *Science and Empires: Historical Studies about Scientific Development and European Expansion* (Boston: Kluwer Academic, 1992), pp. 79–88. See also Michael Lackner, Iwo Amelung, and Joachim Kurtz, eds., *New Terms for New Ideas: Western Knowledge and Lexical Change in Late Imperial China* (Leiden, Neth.: E. J. Brill, 2001).
4. Nicolas Standaert, "The Investigation of Things and the Fathoming of Principles *(Gewu qiongli)* in the Seventeenth-Century Contact between Jesuits and Chinese Scholars," in John W. Witek, ed., *Ferdinand Verbiest (1622–1688): Jesuit Missionary, Scientist, Engineer and Diplomat* (Nettetal, Ger.: Steyler Verlag, 1994), pp. 395–420.
5. W. L. Idema, "Cannon, Clocks and Clever Monkeys: Europeana, Europeans and Europe in Some Ch'ing Novels," in E. B. Vermeer, ed., *Development and Decline of Fukien Province in the Seventeenth and Eighteenth Centuries* (Leiden, Neth.: E. J. Brill, 1990), p. 468.
6. See Nathan Sivin, "Copernicus in China," in Sivin, *Science in Ancient China: Researches and Reflections* (Brookfield, Vt.: Variorium, 1995), part 4, p. 8.
7. John Barrow, *Travels in China* (London: T. Cadell and W. Davies, 1804), p. 110.

8. Susan Thorne, "'The Conversion of Englishmen and the Conversion of the World Inseparable': Missionary Imperialism and the Language of Class in Early Industrial Britain," in Frederick Cooper and Ann Stoler, eds., *Tensions of Empire: Colonial Cultures in a Bourgeois World* (Berkeley: University of California Press, 1997), pp. 238–262.
9. Federico Masini, *The Formation of Modern Chinese Lexicon and Its Evolution toward a National Language: The Period from 1840 to 1898* (Berkeley: University of California, Berkeley, Journal of Chinese Linguistics Monograph Series, 1993), pp. 15–34. See also Wm. Theodore de Bary and Richard Lufrano, eds., *Sources of Chinese Tradition*, vol. 2 (New York: Columbia University Press, 2000), p. 211.
10. Kume Kunitake, comp., *The Iwakura Embassy 1871–73: A True Account of the Ambassador Extraordinary and Plenipotentiary's Journey of Observation through the United States of America and Europe*, 5 vols. (Matsudo: The Japan Documents, 2002), p. 352.
11. On the inadequacy of China's army and navy, see Ralph Powell, *The Rise of Chinese Military Power* (Princeton, N.J.: Princeton University Press, 1955), pp. 36–50. Compare Richard Smith, "Reflections on the Comparative Study of Modernization in China and Japan: Military Aspects," *Journal of the Hong Kong Branch of the Royal Asiatic Society* 16 (1976): 11–23.
12. Douglas Reynolds, *China, 1898–1912: The Xinzheng Revolution and Japan* (Cambridge: Harvard University Press, 1993), pp. 131–150.
13. Yue Meng, "Hybrid Science versus Modernity: The Practice of the Jiangnan Arsenal, 1864–1897," *East Asian Science, Technology, and Medicine* 16 (1999): 43–45.
14. D. W. Y. Kwok, *Scientism in Chinese Thought, 1900–1950* (New Haven: Yale University Press, 1965).
15. In *Science and Technology in World History: An Introduction* (Baltimore: Johns Hopkins University Press, 1999), pp. 117–140, authors James McClellan and Harold Dorn conclude: "With the arrival of [Matteo] Ricci in China the subsequent history of Chinese science largely becomes its integration into ecumenical, world science." Compare Arnold Pacey, *Technology in World Civilization: A Thousand-Year History* (Cambridge: MIT Press, 1991), pp. 94–97, and Lewis Pyenson and Susan Sheets-Pyenson, *Servants of Nature: A History of Scientific Institutions, Enterprises, and Sensibilities* (New York: Norton, 2000), pp. 382–386.
16. Michael Adas, *Machines as the Measure of Men: Science, Technology, and*

Ideologies of Western Dominance (Ithaca: Cornell University Press, 1989). See also Grant Goodman, *Japan: The Dutch Experience* (London: Athlone Press, 1985), p. 228, and Masao Watanabe, *The Japanese and Western Science* (Philadelphia: University of Pennsylvania Press, 1990).

17. On India, see Bernard Cohn, *Colonialism and Its Form of Knowledge: The British in India* (Chicago: University of Chicago Press, 1996), pp. 5–56. Compare Deepak Kumar, ed., *Science and Empires: Essays in Indian Context, 1700–1947* (New Delhi: Willey Eastern, 1991); and Satpal Sangwan, *Science, Technology and Colonisation: An Indian Experience, 1757–1857* (New Delhi: Anamika Prakashan, 1991).
18. Numata Jirō, *Western Learning: A Short History of the Study of Western Science in Early Modern Japan* (Tokyo: Japan-Netherlands Institute, 1992), pp. 3–4, 97, 147–169.
19. Cohn, *Colonialism and Its Form of Knowledge*, pp. 5–76. See also V. V. Krishna, "The Colonial 'Model' and the Emergence of National Science in India: 1876–1920," in Petitjean, Jami, and Moulin, *Science and Empires*, pp. 57–72.
20. Gyan Prakash, *Another Reason: Science and the Imagination of Modern India* (Princeton, N.J.: Princeton University Press, 1999), pp. 3–14, 17–85. See also Deepak Kumar, "Problems in Science Administration: A Study of the Scientific Surveys in British India," in Petitjean, Jami, and Moulin, *Science and Empires*, pp. 269–280.
21. Derk Bodde, "Prison Life in Eighteenth Century Beijing," *Journal of the American Oriental Society* 89 (April–June 1969): 311–333, gives references to earlier accounts of prisons.
22. Roger Hart, "Beyond Science and Civilization: A Post-Needham Critique," *East Asian Science, Technology, and Medicine* 16 (1999): 88–114.
23. Benjamin Elman, "The Jesuit Role as 'Experts' in High Qing Cartography and Technology," *National Taiwan University History Bulletin* 31 (June 2003): 223–250. See also J. L. Cranmer-Byng and Trevor H. Levere, "A Case Study in Cultural Collision: Scientific Apparatus in the Macartney Embassy to China, 1793," *Annals of Science* 38 (1981): 503–525.
24. James Hevia, *Cherishing Men from Afar: Qing Guest Ritual and the Macartney Embassy of 1793* (Durham, N.C.: Duke University Press, 1995).
25. Peter Winch, "Understanding a Primitive Society," in Bryon Wilson, ed., *Rationality* (Oxford: Basil Blackwell, 1970), pp. 93–102; and Donald F. Lach, *Asia in the Making of Europe*, vol. 2: *A Century of Wonder*, book 3:

The Scholarly Disciplines (Chicago: University of Chicago Press, 1977), p. 395.

26. Fa-ti Fan, *British Naturalists in Qing China: Science, Empire, and Cultural Encounter* (Cambridge: Harvard University Press, 2004), chapter 1.

1. *The Jesuit Legacy*

1. Compare Nicholas Jardine and Emma Spary, "The Natures of Cultural History," in Jardine, J. A. Secord, and E. C. Spary, eds., *Cultures of Natural History* (Cambridge, Eng.: Cambridge University Press, 1996), pp. 3–13.
2. Benjamin Elman, *On Their Own Terms: Science in China, 1550–1900* (Cambridge: Harvard University Press, 2005), chapter 3.
3. Ibid.
4. Christopher Cullen, *Astronomy and Mathematics in Ancient China: The Zhoubi suanjing* (Cambridge, Eng.: Cambridge University Press, 1996), pp. 7–20. See also Nathan Sivin, "State, Cosmos, and Body in the Last Three Centuries B.C.," *HJAS* 55, no. 1 (June 1995): 7.
5. Elman, *On Their Own Terms,* chapter 3.
6. Olaf Pederson, "The Ecclesiastical Calendar and the Life of the Church," in *GRC,* pp. 27–28. Compare Paul Richard Blum, "The Jesuits and the Janus-Faced History of Natural Sciences," in Jurgen Helm and Annette Winkelmann eds., *Religious Confessions and the Sciences in the Sixteenth Century* (Leiden, Neth.: E. J. Brill, 2001), pp. 19–34.
7. See Ugo Baldini, "The Academy of Mathematics of the Collegio Romano from 1553 to 1612," in Mordechai Feingold, ed., *Jesuit Science and the Republic of Letters* (Cambridge: MIT Press, 2002); and Mario Biagioli, "The Social Status of Italian Mathematicians, 1450–1600," *History of Science* 27 (1989): 41–95.
8. Nicolas Standaert, *Handbook of Christianity in China,* vol. 1: *635–1800* (Leiden, Neth.: E. J. Brill, 2001), p. 493. The modern value for a tropical year is 365.2422 mean solar days.
9. Keizō Hashimoto, *Hsü Kuang-ch'i and Astronomical Reform—The Process of the Chinese Acceptance of Western Astronomy, 1629–1635* (Osaka: Kansai University Press, 1988), pp. 16–19.
10. Hashimoto, *Hsü Kuang-ch'i,* pp. 20–25, 41–44; and Peter Engelfriet, *Euclid in China: The Genesis of the First Chinese Translation of Euclid's Elements, Books I–VI* (Leiden, Neth.: E. J. Brill, 1998), p. 344.

11. Nicolas Standaert, "The Classification of Sciences and the Jesuit Mission in Late Ming China," in Jan de Meyer and Peter Engelfriet, eds., *Linked Faiths: Essays on Chinese Religions and Traditional Culture in Honor of Kristofer Schipper* (Leiden, Neth.: E. J. Brill, 2000), pp. 287–317.
12. See Nathan Sivin, "Copernicus in China," in Sivin, *Science in Ancient China: Researches and Reflections* (Brookfield, Vt.: Variorium, 1995), part 4, pp. 1–52.
13. Han Qi, "Astronomy, Chinese and Western: The Influence of Xu Guangqi's Views in the Early and Mid-Qing," in Catherine Jami, Peter Engelfriet, and Gregory Blue, eds., *Statecraft and Intellectual Renewal in Late Ming China: The Cross-Cultural Synthesis of Xu Guangqi (1562–1633)* (Leiden, Neth.: E. J. Brill, 2001), pp. 360–373.
14. Pingyi Chu, "Scientific Dispute in the Imperial Court: The 1664 Calendar Case," *Chinese Science* 14 (1997): 13.
15. Cullen, *Astronomy and Mathematics,* pp. 50–54.
16. Sivin, "Copernicus in China," pp. 14–16.
17. Pingyi Chu, "Trust, Instruments, and Cross-Cultural Scientific Exchanges: Chinese Debate over the Shape of the Earth, 1600–1800," *Science in Context* 12, no. 3 (1999): 395.
18. See Han Qi, "Sino-British Relations through Jesuits in the Seventeenth and Eighteenth Centuries," in Michel Cartier, ed., *La Chine entre amour et haine* (Paris: Institut Ricci, Desclée de Brouwer, 1998), p. 57. See also Tsuen-hsuin Tsien, "Western Impact on China through Translation," *Far Eastern Quarterly* 13 (1954): 310, and Li Yan and Du Shiran, *Chinese Mathematics: A Concise History,* trans. John Crossley and Anthony Lun (Oxford: Clarendon Press, 1987), pp. 208–209.
19. Hashimoto, *Hsü Kuang-ch'i,* pp. 186–189, 212–216, and *SCC,* vol. 3, pp. 444–445.
20. Joseph Needham, *Clerks and Craftsmen in China and the West* (Cambridge, Eng.: Cambridge University Press, 1970), p. 205. See also Jonathan Spence, *Emperor of China: Self-Portrait of K'ang-hsi* (New York: Vintage Books, 1975), p. 68.
21. P. M. Engelfriet, *Euclid in China* (Leiden, Neth.: E. J. Brill, 1998), pp. 53–55. For Aristotle, natural motion occurs, for example, because a rock's natural place is Earth, and thus it falls toward Earth. Violent motion requires a continuous external motive force.
22. Florence Hsia, "Some Observations on the *Observations:* The Decline of

the French Jesuit Scientific Mission in China," *Revue de synthèse* 2–3 (1999): 305–333.

23. Allan Chapman, "Tycho Brahe in China: The Jesuit Mission to Peking and the Iconography of European Instrument-Making Processes," *Annals of Science* 41 (1984): 417–443. See also Zhang Baicun, "The Introduction of European Astronomical Instruments and the Related Technology into China during the Seventeenth Century," *East Asian Science, Technology, and Medicine* 20 (2003): 99–131.
24. Lorraine Daston, "Marvelous Facts and Miraculous Evidence in Early Modern Europe," *Critical Inquiry* 18 (1991): 93–124.
25. Robert Wardy, *Aristotle in China: Language, Categories and Translation* (Cambridge, Eng.: Cambridge University Press, 2000), pp. 76, 136, and *ECCP*, pp. 807–809.
26. Pan Jixing, "The Spread of Georgius Agricola's *De Re Metallica* in Late Ming China," *T'oung Pao* 57 (1991): 108–118. See also Pamela Long, "Of Mining, Smelting, and Printing," *Technology and Culture* 44, no. 1 (January 2003): 97–101.
27. Peter Engelfriet, "The Chinese Euclid and Its European Context," in Catherine Jami and Hubert Delahaye, eds., *L'Europe en Chine: Interactions scientifiques, religieuses et culturelles aux XVIIe et XVIIIe siècles* (Paris: Institut des Hautes Études Chinoises, Collège de France, 1993), pp. 114–118.
28. Aihe Wang, *Cosmology and Political Culture in Early China* (Cambridge, Eng.: Cambridge University Press, 2000), pp. 81–128. Despite the common Western view, the "phases" were not elements. See Geoffrey Lloyd and Nathan Sivin, *The Way and the Word: Science and Medicine in Early China and Greece* (New Haven: Yale University Press, 2002), pp. 253–266.
29. Qiong Zhang, "Demystifying *Qi:* The Politics of Cultural Translation and Interpretation in the Early Jesuit Mission to China," in Lydia Liu, ed., *Tokens of Exchange: The Problem of Translation in Global Circulation* (Durham, N.C.: Duke University Press, 1999), pp. 74–106.
30. Vagnoni, *Kongji gezhi* (Treatise on the composition of the universe), reproduction of the 1633 version in *Tianzhu jiao dongchuan wenxian sanbian* (Taipei: Wenjin chuban she, 1979), A.1b–2b.
31. Ibid., A.4a–6a.
32. Willard Peterson, "Western Natural Philosophy Published in Late Ming China," *Proceedings of the American Philosophical Society* 117, no. 4 (August 1973): 295–322.

33. Christopher Cullen, "The Science/Technology Interface in Seventeenth-Century China: Song Yingxing on *qi* and the *wuxing*," *Bulletin of the School of Oriental and African Studies* 53, no. 2 (1990): 295–318.
34. Kenneth Ch'en, "Matteo Ricci's Contribution to, and Influence on, Geographical Knowledge in China," *Journal of the American Oriental Society* 59, no. 3 (September 1939): 325–359.
35. Ibid., pp. 343–347, and *DMB*, pp. 83–84.
36. *The Library of Philip Robinson*, part 2 (London: Sotheby's Auction Catalogue, 1988), pp. 77–95. See also John Day, "The Search for the Origins of the Chinese Manuscript Copies of Matteo Ricci's Maps," *Imago Mundi* 47 (1995).
37. Richard Smith, "Mapping China's World: Cultural Cartography in Late Imperial Times," in Yeh Wen-hsin, ed., *Landscape, Culture and Power in Chinese Society* (Berkeley: University of California, Berkeley, Center for East Asian Studies, 1998), pp. 71–83.
38. Cordell Yee, "Traditional Chinese Cartography and the Myth of Westernization," in J. B. Harley and David Woodward, eds., *The History of Cartography*, vol. 2, book 2: *Cartography in the Traditional East and Southeast Asian Societies* (Chicago: University of Chicago Press, 1994), pp. 170–202.
39. Chen Minsun, "Ferdinand Verbiest and the Geographical Works by Jesuits in Chinese, 1584–1674," pp. 123–164, and Lin Tongyang, "Ferdinand Verbiest's Contribution to Chinese Geography and Cartography," pp. 135–164, both in John Witek, ed., *Ferdinand Verbiest (1623–1688): Jesuit Missionary, Scientist, Engineer and Diplomat* (Nettetal, Ger.: Steyler Verlag, 1994).
40. James A. Millward, *Beyond the Pass: Economy, Ethnicity, and Empire in Qing Central Asia, 1759–1864* (Stanford: Stanford University Press, 1998), pp. 70–72.
41. James S. Cummins, "Two Missionary Methods in China: Mendicants and Jesuits," *Archivo ibero-americano* 37 (1978): 33–108; and John D. Young, "Chinese Views of Rites and the Rites Controversy, Eighteenth–Twentieth Centuries," in D. E. Mungello, ed., *The Chinese Rites Controversy: Its History and Meaning* (Nettetal, Ger.: Steyler Verlag, 1994), p. 86.
42. Edward Malatesta, S. J., "A Fatal Clash of Wills: The Condemnation of the Chinese Rites by the Papal Legate Carlo Tommaso Maillard de Tourrnon," in Mungello, *Chinese Rites Controversy*, pp. 211–245.
43. Paul Rule, "Towards a History of the Chinese Rites Controversy," in

Mungello, *Chinese Rites Controversy,* pp. 249–266. See also D. E. Mungello, "An Introduction to the Chinese Rites Controversy," in Mungello, *The Chinese Rites Controversy,* pp. 7–8, and *Library of Philip Robinson,* part 2, pp. 31, 98.

44. *Library of Philip Robinson,* part 2, pp. 7, 14. The Robinson Library, before its sale in 1988, included important portions of the Clermont collection. On the Jesuit mission's collapse, see M. Hay, *Failure in the Far East* (London: Neville Spearman, 1956).

2. Recovering the Chinese Classics

1. Benjamin A. Elman, *From Philosophy to Philology: Intellectual and Social Aspects of Change in Late Imperial China,* 2d ed. (Los Angeles: UCLA Asia Institute Monograph Series, 2001), pp. 72–122.
2. See Noel M. Swerdlow, "The Recovery of the Exact Science of Antiquity," in Anthony Grafton, ed., *Rome Reborn: The Vatican Library and Renaissance Culture* (Washington, D.C.: Library of Congress, 1993), pp. 125–168.
3. On the literati-physician, see Robert Hymes, "Not Quite Gentlemen? Doctors in Sung and Yuan," *Chinese Science* 7 (1986): 11–85. See also Nathan Sivin, "Text and Experience in Classical Chinese Medicine," in Don Bates, ed., *Knowledge and the Scholarly Medical Traditions* (Cambridge, Eng.: Cambridge University Press, 1995), pp. 177–204.
4. Catherine Jami, "Imperial Control and Western Learning: The Kangxi Emperor's Performance," *Late Imperial China* 23, no. 1 (June 2002): 28–49.
5. See Catherine Jami, "Learning Mathematical Sciences during the Early and Mid-Ch'ing," in Benjamin A. Elman and Alexander Woodside, eds., *Education and Society in Late Imperial China, 1600–1900* (Berkeley: University of California Press, 1994), pp. 231, 238–240.
6. Jean-Claude Martzloff, *A History of Chinese Mathematics,* trans. Stephen Wilson (New York: Springer-Verlag, 1997), pp. 163–166.
7. *SCC,* vol. 3, pp. 141–145. Compare Catherine Jami, "Western Influence and Chinese Tradition in an Eighteenth-Century Chinese Mathematical Work," *Historia Mathematica* 15 (1988): 311–331; and Peter Engelfriet, *Euclid in China: The Genesis of the First Chinese Translation of Euclid's Elements, Books I–VI* (Leiden, Neth.: E. J. Brill, 1998), p. 438.
8. Nathan, "Copernicus in China," in Sivin, *Science in Ancient China: Researches and Reflections* (Brookfield, Vt.: Variorium, 1995), part 4.

9. Martzloff, *A History of Chinese Mathematics,* pp. 218–219. See also *ECCP,* pp. 473–475.
10. Roger Hahn, *The Anatomy of a Scientific Institution: The Paris Academy of Sciences, 1666–1803* (Berkeley: University of California Press, 1971), pp. 275–285.
11. Limin Bai, "Mathematical Study and Intellectual Transition in the Early and Mid-Qing," *Late Imperial China* 16, no. 2 (December 1995): 23–61. Compare Mario Biagioli, *Galileo Courtier: The Practice of Science in an Age of Absolutism* (Chicago: University of Chicago Press, 1993), pp. 54–59.
12. Mark Elliot, *The Manchu Way: The Eight Banners and Ethnic Identity in Late Imperial China* (Stanford: Stanford University Press, 2001).
13. Li Yan and Du Shiran, *Chinese Mathematics: A Concise History,* trans. John Crossley and Anthony Lun (Oxford, Eng.: Clarendon Press, 1987), pp. 214–216. See also Martzloff, *A History of Chinese Mathematics,* pp. 25, 80.
14. Engelfriet, *Euclid in China,* p. 430.
15. See Andrew C. West, *Catalogue of the Morrison Collection of Chinese Books* (London: School of Oriental and African Studies, 1998), pp. 108, 124–126.
16. Li and Du, *Chinese Mathematics,* pp. 214–216. See also Margaret Baron, *The Origins of the Infinitesimal Calculus* (New York: Dover Publications, 1969), pp. 1–9.
17. Martzloff, *A History of Chinese Mathematics,* p. 119; Engelfriet, *Euclid in China,* p. 432; and Wann-sheng Horng, "Chinese Mathematics at the Turn of the Nineteenth Century: Jiao Xun, Wang Lai, and Li Rui," in Cheng-hung Lin and Daiwie Fu, eds., *Philosophy and Conceptual History of Science in Taiwan* (Dordrecht, Neth.: Kluwer Academic, 1993), p. 175.
18. Kenneth Swope, "Crouching Tigers, Secret Weapons: Military Technology Employed in the Japanese Invasion of Korea, 1592–1598," *Journal of Military History* 69 (2005): 11–42.
19. John Shepherd, *Statecraft and Political Economy on the Taiwan Frontier* (Stanford: Stanford University Press, 1993), pp. 47–90; and Cordell Yee, "A Cartography of Introspection: Chinese Maps as Other Than European," *Asian Art* 5 (1992): 29–45.
20. Cheng K'o-ch'eng, "Cheng Ch'eng-kung's Maritime Expansion and Early Ch'ing Coastal Prohibition," in E. B. Vereer, ed., *Development and Decline of Fukien Province in the Seventeenth and Eighteenth Centuries* (Leiden, Neth.: E. J. Brill, 1990), pp. 228–244.
21. John Wills Jr., "Maritime China from Wang Chih to Shih Lang," in Jonathan Spence and John Wills, eds., *From Ming to Ch'ing: Conquest, Region,*

and Continuity in Seventeenth-Century China (New Haven: Yale University Press, 1979), pp. 228–234.

22. Peter Perdue, *China Marches West: The Qing Conquest of Central Asia* (Cambridge: Harvard University Press, 2005).
23. Kenneth Ch'en, "Matteo Ricci's Contribution to, and Influence on, Geographical Knowledge in China," *Journal of the American Oriental Society* 59, no. 3 (September 1939): 347–359; and *ECCP*, pp. 762–767.
24. *ECCP*, pp. 522, 909.
25. Chen Minsun, "Ferdinand Verbiest and the Geographical Works by Jesuits in Chinese, 1584–1674," in John Witek, ed., *Ferdinand Verbiest (1623–1688): Jesuit Missionary, Scientist, Engineer and Diplomat* (Nettetal, Ger.: Steyler Verlag, 1994), pp. 129–131.
26. Compare Nathan Sivin, "On the Limits of Empirical Knowledge in the Traditional Chinese Sciences," in J. T. Fraser, N. Lawrence, and F. C. Haber, eds., *Time, Science, and Society in China and the West* (Amherst: University of Massachusetts Press, 1986), p. 163.
27. Hilde DeWeerdt, "Regional Descriptions: Administrative and Scholarly Traditions," in Patrick Hanan, ed., *Treasures of the Yenching* (Cambridge: Harvard-Yenching Library, 2003), pp. 139–142, and Michael G. Chang, "Fathoming Qianlong: Imperial Activism, the Southern Tours, and the Politics of Water Control, 1736–1765," *Late Imperial China* 24, no. 2 (2003): 51–108.
28. John Henderson, *The Development and Decline of Chinese Cosmology* (New York: Columbia University Press, 1984).
29. *ECCP*, p. 421. See also Henderson, *Development and Decline of Chinese Cosmology*, pp. 231–253, and Nathan Sivin, "Wang Hsi-shan," in Sivin, *Science in Ancient China*, part 5, pp. 1–27.
30. Mark Elliot, "The Limits of Tartary: Manchuria in Imperial and National Geographies," *JAS* 59, no. 3 (August 2000): 603–646.
31. *Library of Philip Robinson*, part 2, p. 95. Compare Philippe Forêt, *Mapping Chengde: The Qing Landscape Enterprise* (Honolulu: University of Hawaii Press, 2000), chapter 6.
32. Peter Perdue, "Boundaries, Maps, and Movement: Chinese, Russian, and Mongolian Empires in Early Modern Central Eurasia," *International History Review* 20, no. 2 (June 1998): 267–268; and Laura Hostetler, *Qing Colonial Enterprise: Ethnography and Cartography in Early Modern China* (Chicago: University of Chicago Press, 2001), pp. 66–71.
33. Compare Elliot, "Limits of Tartary," pp. 621–632.

34. Perdue, "Boundaries, Maps, and Movement," pp. 263–286; and Joanna Waley-Cohen, "Commemorating War in Eighteenth-Century China," *Modern Asian Studies* 30, no. 4 (1996): 869–899.
35. Laura Hostetler, "Qing Connections to the Early Modern World: Ethnography and Cartography in Eighteenth-Century China," *Modern Asian Studies* 34, no. 3 (2000): 623–662.
36. Compare David Turnbull, "Cartography and Science in Early Modern Europe: Mapping and the Construction of Knowledge Spaces," *Imago Mundi* 48 (1996): 5–24.
37. Hostetler, *Qing Colonial Enterprise,* pp. 71–75. See also James Millward, "'Coming onto the Map': Western Regions' Geography and Cartographic Nomenclature in the Making of the Chinese Empire in Xinjiang," *Late Imperial China* 20, no. 2 (December 1999): 61–98.
38. Hostetler, *Qing Colonial Enterprise,* pp. 74–79; and Elliot, "Limits of Tartary," p. 626.
39. Thomas Barfield, *The Perilous Frontier: Nomadic Empires and China, 221 B.C. to A.D. 1757* (Cambridge: Barfield Publishers, 1989), pp. 277–294.
40. Mark Mancall, *Russia and China: Their Diplomatic Relations to 1728* (Cambridge: Harvard University Press, 1971), pp. 149–159, 209–210, 280–283. See also Eric Widmer, *The Russian Ecclesiastical Mission in Peking during the Eighteenth Century* (Cambridge: Harvard University East Asian Research Center, 1976), pp. 45–58, 174–178.
41. Manfred Porkert, *The Theoretical Foundations of Chinese Medicine: Systems of Correspondence* (Cambridge: MIT Press, 1974), pp. 197–216; and Asaf Goldschmidt, "Changing Standards: Tracing Changes in Acu-moxa Therapy during the Transition from the Tang to the Song Dynasties," *East Asian Science, Technology, and Medicine* 18 (2001): 91–92. See also Nathan Sivin, *Traditional Medicine in Contemporary China* (Ann Arbor: Center for Chinese Studies, University of Michigan, 1987), pp. 249–272.
42. Donald Harper, *Early Chinese Medical Literature: The Mawangdui Medical Manuscripts* (London: Kegan Paul, 1998).
43. Shigehisa Kuriyama, *The Expressiveness of the Body and the Divergence of Greek and Chinese Medicine* (New York: Zone Books, 1999), pp. 237–242, 251–259.
44. Porkert, *Theoretical Foundations of Chinese Medicine,* p. 43.
45. Charlotte Furth, *A Flourishing Yin: Gender in China's Medical History, 960–1665* (Berkeley: University of California Press, 1999), pp. 135–137.

46. Wu Yiyi, "A Medical Line of Many Masters: A Prosopographical Study of Liu Wansu and His Disciples from the Jin to the Early Ming," *Chinese Science* 11 (1993–1994): 47.
47. Helen Dunstan, "The Late Ming Epidemics: A Preliminary Survey," *Ch'ing-shih wen-t'i* 3, no. 3 (November 1975): 1–65. See also Marta Hanson, "Robust Northerners and Delicate Southerners: The Nineteenth-Century Invention of a Southern Medical Tradition," *positions* 6, no. 3 (Winter 1998): 548, n. 43.
48. Joanna Grant, *A Chinese Physician: Wang Ji and the 'Stone Mountain Medical Case Histories'* (London: Routledge Curzon, 2003).
49. Chao Yuan-ling, "Medicine and Society in Late Imperial China: A Study of Physicians in Suzhou," Ph.D. diss., University of California, Los Angeles, 1995, pp. 63–68.
50. Catherine Despeux, "The System of the Five Circulatory Phases and the Six Seasonal Influences *(wuyun liuqi)*, a Source of Innovation in Medicine under the Song (960–1279)," trans. Janet Lloyd, in Elisabeth Hsu, ed., *Innovation in Chinese Medicine* (Cambridge, Eng.: Cambridge University Press, 2001), pp. 121–165.
51. Sivin, *Traditional Medicine,* pp. 460–461.
52. *ECCP,* pp. 322–324.
53. Marta Hanson, "The *Golden Mirror* in the Imperial Court of the Qianlong Emperor, 1739–1743," *Early Science and Medicine* 8, no. 2 (2003): 112–147.
54. Chao, "Medicine and Society," pp. 66–70, 97; and *SCC,* vol. 6, part 6, p. 26.
55. Mingjie Hu, "Merging Chinese and Western Mathematics: The Introduction of Algebra and the Calculus in China, 1859–1903," Ph.D. diss., Princeton University, 1998, p. 288. See also Frank Swetz, trans., *The Sea Island Mathematical Manual: Surveying and Mathematics in Ancient China* (University Park: Pennsylvania State University Press, 1992), pp. 7–16.
56. Benjamin A. Elman, *On Their Own Terms: Science in China, 1550–1900* (Cambridge: Harvard University Press, 2005), chapter 6.
57. Martzloff, *A History of Chinese Mathematics,* pp. 32–33.
58. Shigeru Jochi, "The Influence of Chinese Mathematical Texts on Seki Kowa," Ph.D. diss., London University, 1993, pp. 203–204. Andrea Bréard, "On Mathematical Terminology: Culture Crossing in Nineteenth-Century China," in Michael Lackner, Iwo Amelung, and Joachim Kurtz, eds., *New Terms for New Ideas: Western Knowledge and Lexical Change in Late Imperial China* (Leiden, Neth.: E. J. Brill, 2001), pp. 318–320.

59. Li and Du, *Chinese Mathematics,* pp. 225–226.
60. Siu Man-Keung and Alexeï Volkov, "Official Curriculum in Traditional Chinese Mathematics: How Did Candidates Pass the Examinations?" *Historia Scientiarum* 9, no. 1 (1999): 85–99.
61. Li and Du, *Chinese Mathematics,* pp. 226–227.
62. Martzloff, *A History of Chinese Mathematics,* pp. 149–152. Compare Lay-Yong Lam, *A Critical Study of the Yang Hui Suan Fa, a Thirteenth-Century Mathematical Treatise* (Singapore: Singapore University Press, 1977), pp. 36–39.
63. Ulrich Libbrecht, *Chinese Mathematics in the Thirteenth Century: The Shu-shu Chiu-chang of Ch'in Chiu-shao* (Cambridge: MIT Press, 1973).
64. Li and Du, *Chinese Mathematics,* pp. 110–114, 231; and Martzloff, *A History of Chinese Mathematics,* pp. 143–149. See also Libbrecht, *Chinese Mathematics,* p. 18.
65. Tatsuhiko Kobayashi, "What Kind of Mathematics and Terminology Was Transmitted into Eighteenth-Century Japan from China?" *Historia Scientiarum* 12, no. 2 (2002): 2–3.
66. Ken'ichi Sato, "Reevaluation of *Tengenjutsu* or *Tianyuanshu:* In the Context of Comparison between China and Japan," *Historia Scientiarum* 5, no. 1 (1995): 59–60.
67. See Jock Hoe, "Zhu Shijie and His *Jade Mirror of the Four Unknowns,*" in *First Australian Conference on the History of Mathematics: Proceedings of a Conference at Monash University* (Clayton, Aust.), nos. 6 and 7, November 1980, p. 105.
68. Martzloff, *A History of Chinese Mathematics,* pp. 153–157.
69. Alexander Wylie, *Notes on Chinese Literature* (Shanghai: American Presbyterian Mission Press, 1867), pp. 115–116; and Li and Du, *Chinese Mathematics,* pp. 242, 251.
70. Li and Du, *Chinese Mathematics,* pp. 232–233, and Paul Cohen, *Between Tradition and Modernity: Wang T'ao and Reform in Late Ch'ing China* (Cambridge: Harvard University Council on East Asian Studies, 1987), pp. 176–177.
71. Sivin, "Copernicus in China," pp. 45–50. See also Martzloff, *A History of Chinese Mathematics,* pp. 166–172.
72. Ruan Yuan, "Chouren zhuan fanli" (Outline of the *CRZ*), in *CRZ,* pp. 1–5. See also Pingyi Chu, "Remembering Our Grand Tradition: The Historical Memory of the Scientific Exchanges between China and Europe," *History of Science* 41 (2003): 194–199.

73. Limin Bai, "Mathematical Study and Intellectual Transition," pp. 23–61.
74. Tian Miao, "*Jiegenfang, Tianyuan,* and Algebra in Qing China," *Historia Scientiarum* 9, no. 1 (1999): 101–119.
75. Wylie, *Notes on Chinese Literature,* pp. 122–123; Li and Du, *Chinese Mathematics,* pp. 234–240 and 254; and Martzloff, *A History of Chinese Mathematics,* p. 358. See also Elias Loomis, *Elements of Analytical Geometry and of the Integral Calculus* (New York: Harper and Brothers, 1851).
76. See *CRZ,* 30.366. See also Wylie, *Notes on Chinese Literature,* pp. 124–125, and *ECCP,* p. 144.

3. The Rise of Imperial Chinese Manufacturing and Trade

1. Catherine Pagani, *"Eastern Magnificence and European Ingenuity": Clocks of Late Imperial China* (Ann Arbor: University of Michigan Press, 2001), pp. 39–57, 70–74. On earlier clockwork in China, see *SSC,* vol. 4, pp. 220–266.
2. Catherine Jami, "Western Devices for Measuring Time and Space: Clocks and Euclidian Geometry in Late Ming and Ch'ing China," in Chun-chieh Huang and Erik Zürcher, eds., *Time and Space in Chinese Culture* (Leiden, Neth.: E. J. Brill, 1995), pp. 169–200.
3. Pagani, *"Eastern Magnificence,"* pp. 26–57, 181–184.
4. Ibid., pp. 58–98.
5. Ibid., pp. 76–78, 91–93. See also W. L. Idema, "Cannon, Clocks and Clever Monkeys: Europeana, Europeans and Europe in Some Ch'ing Novels," in E. B. Vermeer, ed., *Development and Decline of Fukien Province in the Seventeenth and Eighteenth Centuries* (Leiden, Neth.: E. J. Brill, 1990), pp. 459–488.
6. Larry Stewart, *The Rise of Public Science: Rhetoric, Technology, and Natural Philosophy in Newtonian Britain* (Cambridge, Eng.: Cambridge University Press, 1992), pp. 183–211.
7. Michael Mahoney, "Charting the Globe and Tracking the Heavens: Navigation and the Sciences in the Early Modern Era," in Brett Steele and Tamera Dorland, eds., *The Heirs of Archimedes: Technology, Science and the Art of Warfare through the Age of Enlightenment* (Cambridge: MIT Press, 2005).
8. Emily Curtis, "Plan of the Emperor's Glassworks," *Arts Asiatiques* (Paris) 56 (2001): 81–90.
9. Yang Boda, "An Account of Qing Dynasty Glassmaking," in *Scientific Re-*

search in Early Chinese Glass (Corning, N.Y.: Corning Museum of Glass, 1991), p. 144.

10. Zhang Rong, "Imperial Glass of the Yongzheng Reign," in *Elegance and Radiance: Grandeur in Qing Glass, The Arthur K. F. Lee Collection* (Hong Kong: Art Museum, Chinese University of Hong Kong, 2000), p. 64. See also Rosemary Scott, "Eighteenth Century Overglaze Enamels: The Influence of Technological Development on Painting Style," in Scott and Graham Hutt, eds., *Colloquies on Art and Archaeology in Asia* (London: Percival David Foundation of Chinese Art, 1987), pp. 156–158.
11. Zhang, "Imperial Glass of the Yongzheng Reign," p. 63.
12. Cary Liu, "Architects and Builders in the Qing Dynasty Yuanming Yuan Imperial Garden-Palace," *Hong Kong University Museum Journal* 1 (2002): 38–59.
13. Wong Young-tsu, *A Paradise Lost: The Imperial Garden Yuanming Yuan* (Honolulu: University of Hawaii Press, 2001), pp. 59–65.
14. George Loehr, "The Sinicization of Missionary Artists and Their Works at the Manchu Court during the Eighteenth Century, *Cahiers d'histoire mondiale* 8 (1963): 795–803. See also Crossley, *A Translucent Mirror: History and Identity in Qing Imperial Ideology* (Berkeley: University of California Press, 1999).
15. John Ayers and Rose Kerr, *Blanc de Chine: Porcelain from Dehua* (Chicago: Art Media Resources, 2002); Chuimei Ho, "The Ceramic Trade in Asia, 1602–82," in A. J. H. Latham and Heita Kawakatsu, eds., *Japanese Industrialization and the Asian Economy* (London: Routledge, 1994), p. 39.
16. Craig Clunas, "The Cost of Ceramics and the Cost of Collecting Ceramics in the Ming Period," *Bulletin of the Oriental Ceramic Society of Hong Kong* 8 (1986–1988): 47–53.
17. Stephen Little, *Chinese Ceramics of the Transitional Period: 1620–1683* (New York: China Institute in America, 1983), pp. 1–28; and Patricia Ferguson, *Cobalt Treasures: The Bell Collection of Chinese Blue and White Porcelain* (Toronto: Gardiner Museum of Ceramic Art, 2003).
18. Ayers and Kerr, *Blanc de Chine.* See also J. H. Plumb, *In the Light of History* (Boston: Houghton Mifflin, 1973), pp. 58–59.
19. See Elisabetta Corsi, "Nian Xiyao's (1671–1738) Rendering of Western Perspective in the Prologues to 'Science of Vision,'" in Antonio Forte and Federico Masini, eds., *A Life Journey to the East: Sinological Studies in Memory of Guliano Bertuccioli* (Kyoto: Scuola italiana di studi sull' Asia Orientale, 2002), pp. 201–233.

20. Michel Beurdeley and Guy Raindre, *Qing Porcelain* (New York: Rizzoli, 1986), pp. 86, 92–95, 112–115, 124–132, 136–137, and 141–142; and Helmut Brinker and Albert Lutz, *Chinese Cloisonné: The Pierre Uldry Collection,* trans. Susanna Swoboda (New York: Asia Society Galleries, 1989), p. 23.
21. Rose Kerr and Nigel Wood, *SCC,* vol. 5, part 12 (2004), pp. 709–798. See also Lydia Liu, "Robinson Crusoe's Earthenware Pot," *Critical Inquiry* 25 (Summer 1999): 749–750.
22. Compare Karl Berling, ed., *Meissen China: An Illustrated History* (New York: Dover Books, 1972), pp. 1–6; and Cyril Stanley Smith, "Porcelain and Plutonism," in Smith, *A Search for Structure: Selected Essays on Science, Art, and History* (Cambridge: MIT Press, 1981), pp. 174–190.
23. Beurdeley and Raindre, *Qing Porcelain,* pp. 32–34.
24. Ibid., pp. 35–37, 143–144, 192–210. See also Martin Schonfeld, "Was There a Western Inventor of Porcelain?" *Technology and Culture* 39, no. 4 (1998): 716–727.
25. Kwang Tsing Wu, "Ming Printing and Printers," *HJAS* 7 (1943): 203–260. See also Peter J. Golas, "Technical Representation in China: Tools and Techniques of the Trade." *East Asian Science, Technology, and Medicine* 20 (2003): 11–44.
26. Robert E. Hegel, *Reading Illustrated Fiction in Late Imperial China* (Stanford: Stanford University Press, 1998), pp. 250–289.
27. Cynthia J. Brokaw, "On the History of the Book in China," in Brokaw and Kai-wing Chow, eds., *Printing and Book Culture in Late Imperial China* (Berkeley: University of California Press, 2005), pp. 3–54.
28. Lucille Chia, *Printing for Profit: The Commercial Publishing of Jianyang, Fujian (Eleventh to Seventeenth Centuries)* (Cambridge: Harvard University Asia Center, 2002). See also Cynthia Brokaw, "Commercial Publishing in Late Imperial China: The Zou and Ma Family Businesses of Sibao, Fujian," *Late Imperial China* 17, no. 1 (June 1996): 49–92.
29. Benjamin A. Elman, *From Philosophy to Philology: Intellectual and Social Aspects of Change in Late Imperial China,* 2d ed. (Los Angeles: UCLA Asia Institute Monograph Series, 2001), pp. 178–208.
30. Sakai Tadao, "Confucianism and Popular Education Works," in Wm. Theodore de Bary et al., *Self and Society in Ming Thought* (New York: Columbia University Press, 1970), pp. 331–341.
31. Shang Wei, "The Making of the Everyday World: *Jin Ping Mei* and Encyclopedias for Daily Use," in David Wang and Shang Wei, eds., *Dynastic Decline*

and Cultural Innovation: From the Late Ming to the Late Qing and Beyond (Cambridge: Harvard University Asia Center, 2005).

32. Patrick Hanan, *The Chinese Vernacular Story* (Cambridge: Harvard University Press, 1981), pp. 60–61; and Hanan, "The Sources of the *Chin P'ing Mei*," *Asia Major* 10, no. 1 (1963): 60–62. Compare Shang Wei, "*Jin Ping Mei Cihua* and Late Ming Print Culture," in Judith Zeitlin and Lydia Liu, eds., *Writing and Materiality in China: Essays in Honor of Patrick Hanan* (Cambridge: Harvard Asian Monograph Series, 2003), pp. 187–231.
33. *ECCP*, p. 183.
34. Fujitsuka Chikashi, *Shinchō bunka tōden no kenkyū* (Tokyo: Kokusho kankōkai, 1975), pp. 27–44.
35. Gari Ledyard, "Korean Travellers in China over Four Hundred Years, 1488–1887," *Occasional Papers on Korea* (March 1974): 1–42.
36. Benjamin A. Elman, "Qing Learning and *Kōshōgaku* in Tokugawa Japan," in Joshua Fogel, ed., *Sagacious Monks and Bloodthirsty Warriors: Chinese Views of Japan in the Ming-Qing Period* (Norwalk, Conn.: EastBridge Press, 2002), pp. 158–182.
37. Ōba Osamu, *Edo jidai ni okeru Chūgoku bunka juyō no kenkyū* (Tokyo: Dōbōsha, 1984); and Laura Hess, "The Reimportation from Japan to China of the Kong Commentary to the *Classic of Filial Piety*," Ph.D. diss., University of Washington, Seattle, 1994.
38. Ssu-yü Teng and Knight Biggerstaff, *An Annotated Bibliography of Selected Chinese Reference Works* (Cambridge: Harvard-Yenching Monograph, 1971), pp. 94–96.
39. R. Kent Guy, *The Emperor's Four Treasuries: Scholars and the State in the Late Ch'ien-lung Era* (Cambridge: Harvard Council on East Asian Studies, 1987), and *ECCP*, p. 121.
40. Charles Gillispie, *Science and Polity in France at the End of the Old Regime* (Princeton, N.J.: Princeton University Press, 1980).
41. Albin Brian, *La Mesure de l'Etat: Administrateurs et géomètres au XVIIIe siècle* (Paris: Albin Michel, 1994), pp. 112–144, 230–255.
42. Michael Mahoney, "Infinitesimals and Transcendent Relations: The Mathematics of Motion in the Late Seventeenth Century," in David Lindberg and Robert Westman, eds., *Reappraisals of the Scientific Revolution* (Cambridge: Harvard University Press, 1990), chapter 12; and Keith Baker, *Condorcet: From Natural Philosophy to Social Mathematics* (Chicago: University of Chicago Press, 1975), pp. viii–ix, 4–16, 114.
43. A. Rupert Hall, *Philosophers at War: The Quarrel between Newton and*

Leibniz (Cambridge, Eng.: Cambridge University Press, 1980), pp. 13–31, 254–259.

44. Roger Hahn, "Laplace and the Mechanistic Universe," in David Lindberg and Ronald Numbers, eds., *God and Nature: Historical Essays on the Encounter between Christianity and Science* (Berkeley: University of California Press, 1986), pp. 256–276.
45. Margaret Jacob, *Scientific Culture and the Making of the Industrial West* (Oxford: Oxford University Press, 1997), pp. 94, 126.
46. J. A. Bennett, "The Mechanics' Philosophy and the Mechanical Philosophy," *History of Science* 24 (1986): 1–28. See also Stewart, *Rise of Public Science,* pp. 143–182.
47. Jacob, *Scientific Culture,* pp. 110–111.
48. Ibid., pp. 4–9.
49. Maurice Daumas et al., *A History of Technology and Invention: Progress through the Ages,* vol. 3: *The Expansion of Mechanization,* trans. Eileen Hennessy (New York: Crown, 1979), pp. 39–80.
50. Jacob, *Scientific Culture,* pp. 99–115.
51. Gillispie, *Science and Polity in France,* pp. 81–89.
52. R. Rappaport, "Government Patronage of Science in Eighteenth Century France," *History of Science* 8 (1969): 119–136; and Jean G. Dhombres, "French Textbooks in the Sciences, 1750–1850," *History of Education* 13 (1984): 153–161.
53. Ervan Garrison, *A History of Engineering and Technology: Artful Methods* (Boca Raton, Fla.: CRC Press, 1991), pp. 130–139. Compare Daumas et al., *A History of Technology and Invention,* vol. 3, pp. 235–270.
54. Gillispie, *Science and Polity in France,* pp. 479–548. See also Karl Alder, "French Engineers Become Professionals; or, How Meritocracy Made Knowledge Objective," in William Clark, Jan Golinski, and Simon Schaffer, eds., *The Sciences in Enlightened Europe* (Chicago: University of Chicago Press, 1999), pp. 94–125.
55. Randall Dodgen, *Controlling the Dragon: Confucian Engineers and the Yellow River in Late Imperial China* (Honolulu: University of Hawaii Press, 2001).
56. Jean Dhombres, "L'enseignement des mathématiques par la 'méthode révolutionnaire': Les leçons de Laplace à l'Ecole normale de l'an III," *Revue d'histoire des sciences* 33 (1980): 315–348, and Dhombres, *Naissance d'un pouvoir: Sciences et savants en France (1793–1824)* (Paris: Payot, 1989).
57. George Macartney, *An Embassy to China; Being the Journal Kept by Lord*

Macartney during His Embassy to the Emperor Ch'ien-lung, 1793–1794, edited with an introduction and notes by J. L. Cranmer-Byng (London: Longmans, 1962), p. 299. See also James Hevia, "Looting Beijing: 1860, 1900," in Lydia Liu, ed., *Tokens of Exchange: The Problem of Translation in Global Circulations* (Durham, N.C.: Duke University Press, 1999), pp. 192–199.

58. John Barrow, *Travels in China* (London: T. Cadell and W. Davies, 1804), p. 110. See also Cranmer-Byng and Levere, "Case Study in Cultural Collision," pp. 503–525.
59. Macartney, *Embassy to China,* pp. 266, 310–311. Compare J. L. Cranmer-Byng and Trevor H. Levere, "A Case Study in Cultural Collision: Scientific Apparatus in the Macartney Embassy to China, 1793," *Annals of Science* 38 (1981): 515; and Joanna Waley-Cohen, "China and Western Technology in the Late Eighteenth Century," *American Historical Review* 98, no. 5 (1993): 1534.
60. Albert Feuerwerker, "Chinese Economic History in Comparative Perspective," in Paul Ropp, ed., *Heritage of China: Contemporary Perspectives on Chinese Civilization* (Berkeley: University of California Press, 1990), pp. 224–241.

4. *Science and the Protestant Mission*

1. James Chandler, *England in 1819: The Politics of Literary Culture and the Case of Romantic Historicism* (Chicago: University of Chicago Press, 1998).
2. David Porter, "A Peculiar but Uninteresting Nation: China and the Discourse of Commerce in Eighteenth-Century England," *Eighteenth-Century Studies* 33, no. 2 (1999–2000): 181–199.
3. Susan Thorne, *Congregational Missions and the Making of an Imperial Culture in Nineteenth-Century England* (Stanford: Stanford University Press, 1999), pp. 13–14, 23–52. See also Paul Cohen, "Christian Missions and Their Impact to 1900," in Denis C. Twitchett and John K. Fairbank, eds., *Cambridge History of China,* vol. 10: *Late Ch'ing, 1800–1911,* part 1 (Cambridge, Eng.: Cambridge University Press, 1978), pp. 547–550.
4. John Fairbank, ed., *The Missionary Enterprise in China and America* (Cambridge: Harvard University Press, 1974). Compare Jonathan Spence, *To Change China: Western Advisers in China, 1620–1960* (Harmondsworth, Eng.: Penguin Books, 1980).
5. Tsuen-hsuin Tsien, "Western Impact on China through Translation," *Far*

Eastern Quarterly 13 (1954): 313. See also Andrew C. West, *Catalogue of the Morrison Collection of Chinese Books* (London: School of Oriental and African Studies, 1998), pp. 108, 124–126.

6. Ting-yee Kuo and Kwang-Ching Liu, "Self-Strengthening: The Pursuit of Western Technology," in Twitchett and Fairbank, *Cambridge History of China,* vol. 10: *Late Ch'ing, 1800–1911,* part 1, pp. 544–549. See also Britton Rosewell, *The Chinese Periodical Press, 1800–1912* (Shanghai: Kelley & Walsh, Ltd., 1933), pp. 22–29.
7. *ECCP,* p. 504. See also Xiong Yuezhi, "Difficulties in Comprehension and Differences in Expression: Interpreting American Democracy in the Late Qing," *Late Imperial China* 23, no. 1 (June 2002): 4–5.
8. Suzanne Wilson Barnett, "Wei Yuan and Westerners: Notes on the Sources of the *Hai-kuo t'u-chih,*" *Ch'ing-shih wen-t'i* 2, no. 4 (November 1970): 1–20.
9. Kuo and Liu, "Self-Strengthening," pp. 552–553.
10. David Wright, *Translating Science: The Transmission of Western Chemistry into Late Imperial China, 1840–1900* (Leiden, Neth.: E. J. Brill, 2000), pp. 86–93; and Bridie Andrews, "The Making of Modern Chinese Medicine, 1895–1937," Ph.D. diss., Cambridge University, 1996, pp. 31–32.
11. Patrick Hanan, "Chinese Christian Literature: The Writing Process," in Hanan, ed., *Treasures of the Yenching* (Cambridge: Harvard-Yenching Library, 2003), pp. 261–283. See also Knight Biggerstaff, *The Earliest Modern Government Schools in China* (Ithaca: Cornell University Press, 1961), pp. 175–176; and David Wright, "The Translation of Modern Western Science in Nineteenth-Century China, 1840–1895," *Isis* 89 (1998): 660.
12. Hugh Shapiro, "How Different Are Western and Chinese Medicine? The Case of Nerves," in Helaine Selin, ed., *Medicine across Cultures: The History of Non-Western Medicine* (Boston: Kluwer Academic, 2003), p. 10.
13. Compare William Lockhart, *The Medical Missionary in China: A Narrative of Twenty Years' Experience* (London: Hurst and Blackett, 1861), pp. 111–172, and Andrews, "The Making of Modern Chinese Medicine," pp. 32–34.
14. Marta Hanson, "Robust Northerners and Delicate Southerners: The Nineteenth-Century Invention of a Southern Medical Tradition," *positions* 6, no. 3 (Winter 1998): 532–542.
15. See Ruth Rogaski, *Hygienic Modernity: Meanings of Health and Disease in Treaty-port China* (Berkeley: University of California Press, 2004), chapter 2.
16. Yi-li Wu, "God's Uterus: Benjamin Hobson and Missionary 'Midwifery' in

Nineteenth-Century China," presented at the conference "The Disunity of Chinese Science," University of Chicago, May 10–11, 2002, pp. 32–34.

17. Yamada Keiji, "Anatometrics in Ancient China," *Chinese Science* 10 (1991): 39–52. See also Saburō Miyashita, "A Link in the Westward Transmission of Chinese Anatomy in the Later Middle Ages," in Nathan Sivin, ed., *Science and Technology in East Asia* (New York: Science History Publications, 1977), pp. 200–204.
18. Warwick Anderson, "Immunities of Empire: Race, Disease, and the New Tropical Medicine, 1900–1920," *Bulletin of the History of Medicine* 70, no. 1 (1996): 94–118. See also Bridie Andrews, "Tuberculosis and the Assimilation of Germ Theory in China, 1895–1937," *Journal of the History of Medicine and Allied Sciences* 52, no. 1 (1997): 142–143.
19. Douglas Haynes, *Imperial Medicine: Patrick Manson and the Conquest of Tropical Disease* (Philadelphia: University of Pennsylvania Press, 2001), pp. 18–27, 36–41.
20. Compare Michael Worboys, "Germs, Malaria and the Invention of Mansonian Tropical Medicine," in David Arnold, ed., *Warm Climates and Western Medicine: The Emergence of Tropical Medicine, 1500–1900* (Amsterdam: Rodopi, 1996), pp. 181–207.
21. Bridie Andrews, "Tailoring Tradition: The Impact of Modern Medicine on Traditional Chinese Medicine, 1887–1937," in Viviane Alleton and Alexeï Volkov, eds., *Notions et Perceptions du Changement en Chine* (Paris: Collège de France, Institut des hautes études chinoises, 1994), pp. 149–166.
22. Zhao Hongjun, "Chinese versus Western Medicine: A History of Their Relations in the Twentieth Century," *Chinese Science* 10 (1991): 32. See also Hugh Shapiro, "The Puzzle of Spermatorrhea in Republican China," *positions* 6, no. 3 (Winter 1998): 565–571.
23. *SCC*, vol. 6, part 6, and Chia-feng Chang, "Aspects of Smallpox and Its Significance in Chinese History," Ph.D. diss., London University, 1996, pp. 124–168.
24. Zhao, "Chinese versus Western Medicine," pp. 22–23.
25. Andrews, "Tailoring Tradition," pp. 155–159. See also Elizabeth Hsu, "The Reception of Western Medicine in China: Examples from Yunnan," in Patrick Petitjean, Catherine Jami, and Anne Marie Moulin, eds., *Science and Empires: Historical Studies about Scientific Development and European Expansion* (Boston: Kluwer Academic, 1992), pp. 91–94.
26. Andrews, "Tailoring Tradition," pp. 150–151. Compare Yi-li Wu, "Transmitted Secrets: The Doctors of the Lower Yangzi Region and Popular Gyne-

cology in Late Imperial China," Ph.D. diss., Yale University, 1998, pp. 203–239.

27. Andrews, "Tailoring Tradition," p. 151.
28. Wright, *Translating Science*, pp. 36–38.
29. John Fryer, "An Account of the Department for the Translation of Foreign Books at the Kiangnan Arsenal, Shanghai," *North-China Herald*, January 29, 1880, pp. 77–81.
30. Wright, *Translating Science*, pp. 43–46.
31. Paul A. Cohen, *Between Tradition and Modernity: Wang T'ao and Reform in Late Ch'ing China* (Cambridge: Council on East Asian Studies, Harvard University, 1987), pp. 9–13, 45–55.
32. Federico Masini, *The Formation of Modern Chinese Lexicon and Its Evolution toward a National Language: The Period from 1840 to 1898* (Berkeley: University of California, Berkeley, *Journal of Chinese Linguistics* Monograph Series, 1993), pp. 57–60.
33. Ibid., pp. 57–62.
34. Adrian Bennett, *John Fryer: The Introduction of Western Science and Technology into Nineteenth-Century China* (Cambridge: Harvard University East Asian Research Center, 1967), p. 140. See also Crosbie Smith and M. Norton Wise, *Energy and Empire: A Biographical Study of Lord Kelvin* (Cambridge, Eng.: Cambridge University Press, 1989), pp. 153–155.
35. Alexander Wylie, "Preface" (in English), pp. i–ii, in Wylie and Li Shanlan, *Shuxue qimeng* (Shanghai: Mohai shuguan, 1853), and Wylie, "Preface" (in Chinese), pp. 1a–3b, in Wylie and Li Shanlan, *Daishu xue* (Shanghai: Mohai shuguan, 1859).
36. Alexander Wylie, "Preface" (in Chinese), p. 2a, in Wylie and Li Shanlan, *Shuxue* qimeng. See also Mingjie Hu, "Merging Chinese and Western Mathematics: The Introduction of Algebra and the Calculus in China, 1859–1903," Ph.D. diss., Princeton University, 1998, pp. 119–123, 136–143, 153, 352. Compare *DSB*, vol. 14, pp. 35–37; and Alexander Wylie, *Notes on Chinese Literature* (Shanghai: American Presbyterian Mission Press, 1867), p. 125.
37. *LHCT*, vol. 1, no. 1 (1857.1.26): 2a, and vol. 1, no. 2 (1857.2.24): 3b–6b. See also Masini, *Formation of Modern Chinese Lexicon*, pp. 84–88.
38. *LHCT*, vol. 1, no. 1 (1857.1.26): 1a–1b.
39. David Wright, "John Fryer and the Shanghai Polytechnic: Making Space for Science in Nineteenth-Century China," *British Journal for the History of Science* 29 (1996): 12–13. Compare James R. Pusey, *China and Charles*

Darwin (Cambridge: Harvard University Press, 1983), pp. 4–5. See also James Moore, *The Post-Darwinian Controversies: A Study of the Protestant Struggle to Come to Terms with Darwin in Great Britain and America* (Cambridge, Eng.: Cambridge University Press, 1979), pp. 15–16, 329–330, 342–343.

40. Wylie, "Preface" (in English), p. i, in Wylie and Li Shanlan, *Dai weiji shiji* (Shanghai: Mohai shuguan, 1859).
41. See Elaine Koppelman, "The Calculus of Operations and the Rise of Abstract Algebra," *Archive for History of Exact Sciences* 7 (1971): 155–242; and Niccolò Guicciardini, *The Development of Newtonian Calculus in Britain, 1700–1800* (Cambridge, Eng.: Cambridge University Press, 1989), pp. vii, 1–6.
42. See Wylie's "Preface" (in Chinese), p. 2b, in Wylie's and Li's *Dai weiji shiji.* See also Richard Yeo, "William Whewell: A Cambridge Historian and Philosopher of Science," in Peter Harman and Simon Mitton, eds., *Cambridge Scientific Minds* (Cambridge, Eng.: Cambridge University Press, 2002), pp. 51–62.
43. Guicciardini, *The Development of Newtonian Calculus,* pp. 139–142.
44. Ibid., pp. 104, 120.
45. Hu, "Merging Chinese and Western Mathematics," p. 195.
46. Ibid., pp. 181–215. See also Jean-Claude Martzloff, *A History of Chinese Mathematics,* trans. Stephen Wilson (New York: Springer-Verlag, 1997), pp. 379–382; and Andrea Bréard, "On Mathematical Terminology: Culture Crossing in Nineteenth-Century China," in Michael Lackner, Iwo Amelung, and Joachim Kurtz, eds., *New Terms for New Ideas: Western Knowledge and Lexical Change in Late Imperial China* (Leiden, Neth.: E. J. Brill, 2001), pp. 308–309.
47. Tian Miao, "*Jiegenfang, Tianyuan,* and *Daishu:* Algebra in the Qing Dynasty," *Historia Scientiarum* 9, no. 1 (1999). See also, on the Taiping Rebellion, Benjamin A. Elman, *From Philosophy to Philology: Intellectual and Social Aspects of Change in Late Imperial China,* 2d ed. (Los Angeles: UCLA Asia Institute Monograph Series, 2001), pp. 287–290.
48. W. H. Medhurst, "A Reading-Room for China," *North-China Herald,* March 12, 1874, pp. 225–226. Compare Jeffrey Auerbach, *The Great Exhibition of 1851* (New Haven: Yale University Press, 1999).
49. Charles Gillispie, *Genesis and Geology: A Study of the Relations of Scientific Thought, Natural Theology, and Social Opinion in Great Britain, 1790–1850* (Cambridge: Harvard University Press, 1951); and Roy Porter, "Charles

Lyell and the Principles of the History of Geology," *British Journal for the History of Science* 9 (1976): 91–103.

50. See Knight Biggerstaff, "Shanghai Polytechnic Institution and Reading Room: An Attempt to Introduce Western Science and Technology to the Chinese," *Pacific Historical Review* 25 (May 1956): 131–134.
51. Cohen, *Between Tradition and Modernity,* p. 182; Wright, "John Fryer," p. 11, n. 55.
52. Ferdinand Dagenais, *John Fryer's Calendar: Correspondence, Publications, and Miscellaneous Papers with Excerpts and Commentary* (Berkeley: University of California Center for Chinese Studies, 1999), version 3, 1895:13.
53. Biggerstaff, "Shanghai Polytechnic," p. 144.
54. Rudolph Wagner, "The Early Chinese Newspapers and the Chinese Public Sphere," *European Journal of East Asian Studies* 1, no. 1 (2001): 3–13, 25–26.
55. Quoted in Li, "Letters to the Editor," p. 743. See also Dagenais, *John Fryer,* 1891:1; Bennett, *John Fryer,* pp. 50–55; and Wright, "John Fryer," pp. 1–16.
56. Biggerstaff, "Shanghai Polytechnic," pp. 148–149.
57. Li, "Letters to the Editor," pp. 730–731, 762.
58. Biggerstaff, "Shanghai Polytechnic," pp. 144–149.
59. "A Chinese Scientific Journal," in *Scientific American* 34, no. 17 (April 29, 1876): 279.
60. Xu Shou, "*Gezhi huibian* xu," in *GZHB,* vol. 1, pp. 3–4. See also Dagenais, *John Fryer,* 1876:2–3.
61. Rogaski, *Hygienic Modernity,* chapter 4.
62. Biggerstaff, "Shanghai Polytechnic," p. 145.
63. Yi-li Wu, "Introducing the Uterus to Chinese Gynecology: Benjamin Hobson and His *Treatise on Midwifery and Diseases of Children* (Fuying Xinshou), 1858," paper presented at the panel on "Representing Western Medicine in Qing and Republican China," Association for Asian Studies Annual Meeting, Chicago, March 24, 2001, pp. 12–13.
64. *GZHB,* vol. 1, pp. 168–169, 195–196, 433–434.
65. David Reynolds, "Re-Drawing China's Intellectual Map: Nineteenth-Century Chinese Images of Science," *Late Imperial China* 12, no. 1 (June 1991): 27–61.

5. *From Textbooks to Darwin*

1. Mingjie Hu, "Merging Chinese and Western Mathematics: The Introduction of Algebra and the Calculus in China, 1859–1903," Ph.D. diss., Prince-

ton University, 1998, p. 82; and Adrian Bennett, *John Fryer: The Introduction of Western Science and Technology into Nineteenth-Century China* (Cambridge: Harvard University East Asia Research Center, 1967), p. 123. Compare David Wright, *Translating Science: The Transmission of Western Chemistry into Late Imperial China, 1840–1900* (Leiden, Neth.: E. J. Brill, 2000), pp. 266–272.

2. See *GZQM* (1879–1880).
3. Ayano Hiroyuki, "H. E. Roscoe to Owens College, Manchester no seido tenkan," *Kagakushi kenkyū* 8, no. 212 (1999): 214–222, and *DSB*, vol. 11, pp. 536–538. See also Peter Harman and Simon Mitton, eds., *Cambridge Scientific Minds* (Cambridge, Eng.: Cambridge University Press, 2002).
4. Balfour Stewart, *Lessons in Elementary Physics*, new ed. (London: Macmillan, 1878), p. iii; and Ayano Hiroyuki, "Jūkyū seiki koban no Igirisu (England) ni okeru kagaku no kaikaku undō," *Kagakushi kenkyū* 36, no. 204 (1997): 209–217. See also John Fryer, "Science in China," *Nature* 601, no. 24 (May–October 1881): 9–11, 54–57.
5. Knight Biggerstaff, "Shanghai Polytechnic Institution and Reading Room: An Attempt to Introduce Western Science and Technology to the Chinese," *Pacific Historical Review* 25 (May 1956): 127–149. See also David Wright, "John Fryer and the Shanghai Polytechnic: Making Space for Science in Nineteenth-Century China," *British Journal for the History of Science* 29 (1996): 1–16. Compare Roy MacLeod, *The Creed of Science in Victorian England* (Aldershot: Variorum, 2000).
6. Ferdinand Dagenais, *John Fryer's Calendar: Correspondence, Publications, and Miscellaneous Papers with Excerpts and Commentary* (Berkeley: Center for Chinese Studies, University of California, 1999), version 3, 1887:7, 1889:1.
7. *GZHB*, vol. 5, pp. 200–203.
8. Li Hongzhang, "Xu," in *GZQM* (1886), vol. 1, pp. 1a–4b.
9. Dagenais, *John Fryer's Calendar*, 1894:2.
10. David Wright, "The Great Desideratum: Chinese Chemical Nomenclature and the Transmission of Western Chemical Concepts," *Chinese Science* 14 (1997): 47.
11. Compare T. H. Huxley, *Introductory* (for the *Science Primers*) (London: Macmillan, 1880).
12. *GZQM* (1886), vol. 3, pp. 1.1a–33a, 2.34a–53a.
13. Ibid., vol. 3, pp. 3.54a–71a, 4.72a–111a.

14. Ibid., vol. 4, pp. 1.1a–6a, and vol. 5, pp. 1.1a–5a, 5.53a–61a.
15. Ibid., vol. 5, pp. 8.119a–120b.
16. Fa-ti Fan, "Hybrid Discourse and Textual Practice: Sinology and Natural History in the Nineteenth Century," *Historia Scientiarum* 38 (2000): 26–56.
17. *GZQM* (1886), vol. 6, pp. 1a–b, 110b–112b.
18. J. D. Hooker, *Botany* (New York: Appleton and Company, 1877), pp. 100–102.
19. *GZQM* (1886), vol. 6, pp. 1a–b, 110b–112b. See also Frederick Gregory, "The Impact of Darwinian Evolution on Protestant Theology in the Nineteenth Century," in David Lindberg and Ronald Numbers, eds., *God and Nature: Historical Essays on the Encounter between Christianity and Science* (Berkeley: University of California Press, 1986), pp. 220–241.
20. Compare Hooker, *Botany,* p. 101.
21. *GZQM* (1886), vol. 7, pp. 3a–5b, 6a–7a, 39b–79a, 131b. Compare Michael Foster, *A Course of Elementary Practical Physiology,* assisted by J. N. Langley, 3d ed. (London: Macmillan, 1878).
22. *GZQM* (1886), vol. 9, pp. 51b–55a, 55a–58b, 60a–68a.
23. John Fryer, "Chinese Prize Essays: Report of the Chinese Prize Essay Scheme in Connection with the Chinese Polytechnic Institution and Reading Rooms, Shanghai, for 1886 and 1887," *North-China Herald,* January 25, 1888, pp. 100–101.
24. Benjamin Elman, *A Cultural History of Civil Examinations in Late Imperial China* (Berkeley: University of California Press, 2000), pp. 380–420.
25. The lead article "The Progress of Foreign Studies," in the *North-China Herald,* April 14, 1893, pp. 513–514, described some sources for answering the questions. See also Wright, *Translating Science,* pp. 163–173, and Benjamin A. Elman, *On Their Own Terms: Science in China, 1550–1900* (Cambridge: Harvard University Press, 2005), pp. 430–432.
26. John Fryer, "Second Report of the Chinese Prize Essay Scheme in Connection with the Chinese Polytechnic Institution and Reading Rooms, Shanghai, from July 1887 to July 1889," *North-China Herald,* July 20, 1889, p. 86; and Biggerstaff, "Shanghai Polytechnic," p. 149. See also Elman, *On Their Own Terms,* pp. 337–339.
27. Fryer, "Second Report," p. 85.
28. Elman, *A Cultural History,* pp. 594–602, 605–608; and Wright, *Translating Science,* pp. 178–179.
29. Elman, *A Cultural History,* pp. 594–602.

30. *GZSYKY,* vol. 1, 1889, p. 20b. See also Elman, *A Cultural History,* pp. 389–399.
31. *GZSYKY,* vol. 1, 1889, p. 6b.
32. Ting-yee Kuo and Kwang-Ching Liu, "Self-Strengthening: The Pursuit of Western Technology," in Denis C. Twitchett and John K. Fairbank, eds., *Cambridge History of China,* vol. 10: *Late Ch'ing, 1800–1911,* part 1 (Cambridge, Eng.: Cambridge University Press, 1978), pp. 531–532; and Bridie Andrews, "Tailoring Tradition: The Impact of Modern Medicine on Traditional Chinese Medicine, 1887–1937," in Viviane Alleton and Alexeï Volkov, eds., *Notions et Perceptions du Changement en Chine* (Paris: Collège de France, Institut des hautes études chinoises, 1994), p. 152.
33. See Xiaoqun Xu, "'National Essence' vs. 'Science': Chinese Native Physicians' Fight for Legitimacy, 1912–37," *Modern Asian Studies* 31, no. 4 (1997): 848.
34. Howard Boorman and Richard Howard, eds., *Biographical Dictionary of Republican China* (New York: Columbia University Press, 1967), pp. 170–189.
35. Andrews, "Tailoring Tradition," pp. 152–153.
36. *GZSYKY,* vol. 2, 1891, p. 1b (topic), pp. 1a–17a (prize essays).
37. Ibid., 1892, pp. 3a–b (topic), pp. 4b–6b, 33b–34b (prize essays).
38. Ibid., 1893, p. 2a (topic), pp. 4a–9b (prize essays).
39. Elman, *On Their Own Terms,* p. 345.
40. *GZSYKY,* vol. 1, 1889, p. 1a.
41. Ibid., p. 8b.
42. Ibid., p. 12b.
43. *GZHB,* vol. 2, pp. 13–15 (September 1877). Compare Wright, "John Fryer," p. 13, and James R. Pusey, *China and Charles Darwin* (Cambridge: Harvard University Press, 1983), pp. 4–5.
44. See *GZSYKY,* vol. 1, 1889, pp. 21a–b.
45. David Wright, "Yan Fu and the Tasks of the Translator," in Michael Lackner, Iwo Amelung, and Joachim Kurtz, eds., *New Terms for New Ideas: Western Knowledge and Lexical Change in Late Imperial China* (Leiden, Neth.: E. J. Brill, 2001), pp. 235–255.
46. *GZSYKY,* vol. 1, 1889, p. 21a.
47. *North-China Herald,* January 29, 1902, p. 180; September 22, 1905, pp. 697–698; January 25, 1907, p. 202; and February 21, 1908, pp. 418–419. See also Natascha Vittinghoff, "Unity vs. Uniformity: Liang Qichao and the

Invention of a 'New Journalism' for China," *Late Imperial China* 23, no. 1 (June 2002): 91–143.

6. *Government Arsenals Spur New Technologies*

1. Hans J. van de Ven, "War in the Making of Modern China," *Modern Asian Studies* 30, no. 4 (1996): 737–756.
2. David Wright, "The Great Desideratum: Chinese Chemical Nomenclature and the Transmission of Western Chemical Concepts," *Chinese Science* 14 (1997): 35–70.
3. Knight Biggerstaff, *The Earliest Modern Government Schools in China* (Ithaca: Cornell University Press, 1961), pp. 174–175; and Ferdinand Dagenais, *John Fryer's Calendar: Correspondence, Publications, and Miscellaneous Papers with Excerpts and Commentary* (Berkeley: Center for Chinese Studies, University of California, 1999), version 3, 1868:8, 1871:2. See also David Wright, *Translating Science: The Transmission of Western Chemistry into Late Imperial China, 1840–1900* (Leiden, Neth.: E. J. Brill, 2000), pp. 238–240.
4. Jonathan Spence, *To Change China: Western Advisers in China 1620–1960* (Harmondsworth, Eng.: Penguin Books, 1980), pp. 34–56.
5. Biggerstaff, *Earliest Modern Government Schools,* p. 174. See also Adrian Bennett, *John Fryer: The Introduction of Western Science and Technology into Nineteenth-Century China* (Cambridge: Harvard University East Asian Research Center, 1967), pp. 20–21, 149, n. 7.
6. *ECCP,* p. 862.
7. Spence, *To Change China,* pp. 74–75, and James Hevia, "Looting Beijing," in Lydia Liu, ed., *Tokens of Exchange: The Problem of Translation and Interpretation in the Early Jesuit Mission to China* (Durham, N.C.: Duke University Press, 1999), pp. 193–199.
8. Biggerstaff, *Earliest Modern Government Schools,* pp. 6–14. See also *ECCP,* pp. 380–383, and Ssu-yü Teng and John Fairbank, eds., *China's Response to the West* (Cambridge: Harvard University Press, 1954), p. 35.
9. Nancy Evans, "The Canton T'ung-wen Kuan: A Study of the Role of Bannermen in One Area of Self-Strengthening," *Papers on China* (Harvard), 22A (1969): 89–103, and Federico Masini, *The Formation of Modern Chinese Lexicon and Its Evolution toward a National Language: The Period from 1840 to 1898* (Berkeley: *Journal of Chinese Linguistics* Monograph Series, University of California, 1993), pp. 46–53.

10. Biggerstaff, *Earliest Modern Government Schools,* pp. 15–18, and *ECCP,* pp. 402–404, 721, 754. See also Teng and Fairbank, *China's Response to the West,* pp. 61–63.
11. Ting-yee Kuo and Kwang-Ching Liu, "Self-Strengthening: The Pursuit of Western Technology," in Denis C. Twitchett and John K. Fairbank, eds., *Cambridge History of China,* vol. 10: *Late Ch'ing, 1800–1911,* part 1 (Cambridge, Eng.: Cambridge University Press, 1978), pp. 519–521; and Spence, *To Change China,* pp. 108–111.
12. David Pong, *Shen Pao-chen and China's Modernization in the Nineteenth Century* (Cambridge, Eng.: Cambridge University Press, 1994), pp. 134–160, 178. See also Teng and Fairbank, *China's Response to the West,* pp. 81–83.
13. Teng and Fairbank, *China's Response to the West,* pp. 76–77.
14. Biggerstaff, *Earliest Modern Government Schools,* pp. 19–34, and *ECCP,* p. 480.
15. Teng and Fairbank, *China's Response to the West,* pp. 64–65.
16. Mary Wright, *The Last Stand of Chinese Conservatism: The T'ung-chih Restoration, 1862–1874* (Stanford: Stanford University Press, 1962), pp. 211–212. See also Yue Meng, "Hybrid Science versus Modernity: The Practice of the Jiangnan Arsenal, 1864–1897," *East Asian Science, Technology, and Medicine* 16 (1999): 13–52.
17. *ECCP,* pp. 721–722. See also *Scientific American* (June 9, 1866).
18. Masini, *Formation of Modern Chinese Lexicon,* pp. 62–71.
19. Wright, *Translating Science,* pp. 108–109.
20. Bennett, *John Fryer,* p. 50, and Dagenais, *John Fryer's Calendar,* 1867:1, 1872:1. On chemistry, see James Reardon-Anderson, *The Study of Change: Chemistry in China, 1840–1949* (Cambridge, Eng.: Cambridge University Press, 1991), pp. 29–52.
21. Dagenais, *John Fryer's Calendar,* 1886:1–2.
22. John Fryer, "An Account of the Department for the Translation of Foreign Books at the Kiangnan Arsenal, Shanghai," *North-China Herald,* January 29, 1880, pp. 79–81. Compare Wright, *Translating Science,* pp. 229–240. See also Cynthia J. Brokaw, "On the History of the Book in China," in Brokaw and Kai-wing Chow, eds., *Printing and Book Culture in Late Imperial China* (Berkeley: University of California Press, 2005), pp. 3–54.
23. Dagenais, *John Fryer's Calendar,* 1880:5, 1883:1; and Bennett, *John Fryer,* pp. 46, 60–61. See also Adrian Bennett, *Missionary Journalist in China: Young J. Allen and His Magazines, 1860–1883* (Athens: University of Georgia Press, 1983), pp. 294–296.

24. Justus Doolittle, *Vocabulary and Hand-book of the Chinese Language*, 2 vols. (Fuzhou: Rozario, Marcal and Company, 1872), vol. 1, pp. 175–178, 308–318.
25. Masini, *Formation of Modern Chinese Lexicon*, pp. 50–53. See also Anders Lundgren and Bernadette Bensaude-Vincent, ed., *Communicating Chemistry: Textbooks and Their Audiences, 1789–1939* (Canton, Mass.: Science History Publications, 2000).
26. John Fryer, "Scientific Terminology: Present Discrepancies and Means of Securing Uniformity," in *Records of the General Conference of the Protestant Missionaries of China Held at Shanghai, May 7–20, 1890* (Shanghai: American Presbyterian Mission Press, 1890), pp. 531–549. See also Bennett, *John Fryer*, pp. 30–33.
27. Calvin W. Mateer and W. M. Hayes, "Letter to the Committee of Education Association on Terms," and "The Committee on Terminology's Report," both in *John Fryer Papers*, Bancroft Library, University of California, Berkeley. Compare Bennett, *John Fryer*, pp. 32–33. See also Wang Yangzong, "A New Inquiry into the Translations of Chemical Terms by John Fryer and Xu Shou," in Michael Lackner, Iwo Amelung, and Joachim Kurtz, eds., *New Terms for New Ideas: Western Knowledge and Lexical Change in Late Imperial China* (Leiden, Neth.: E. J. Brill, 2001), pp. 274–281.
28. Iwo Amelung, "Weights and Forces: The Reception of Western Mechanics in Late Imperial China," in Lackner, Amelung, and Kurtz, *New Terms for New Ideas*, pp. 198–199.
29. Ibid., pp. 204–207.
30. Calvin Mateer, "The Revised List of the Chemical Elements," *Chinese Recorder* 29 (February 1898): 87–94.
31. Fryer, "Account of the Department for the Translation of Foreign Books," pp. 78–79. See also Bennett, *John Fryer*, pp. 14–17, 20–26, 69; and Biggerstaff, *Earliest Modern Government Schools*, pp. 174–175. Compare Spence, *To Change China*, pp. 156–157.
32. Alexander Wylie, *Notes on Chinese Literature* (Shanghai: American Presbyterian Mission Press, 1867), pp. 118–119, 120.
33. *ECCP*, pp. 240–243, 331–333. Compare Zhang Baichun, "An Inquiry into the History of the Chinese Terms *Jiqi* (Machine) and *Jixie* (Machinery)," in Lackner, Amelung, and Kurtz, *New Terms for Old Ideas*, pp. 177–195.
34. Wang Yangzong, "A New Inquiry into the Translations of Chemical Terms," pp. 282–283.

35. Fryer, "Account of the Department for the Translation of Foreign Books," pp. 78–79, and Bennett, *John Fryer,* pp. 34–35.
36. Biggerstaff, *Earliest Modern Government Schools,* pp. 166–167, 173; and Bennett, *John Fryer,* pp. 18–25.
37. Meng, "Hybrid Science," pp. 32–33. See also Biggerstaff, *Earliest Modern Government Schools,* p. 79.
38. Biggerstaff, *Earliest Modern Government Schools,* pp. 166–171. See also Wann-sheng Horng, "Chinese Mathematics at the Turn of the Nineteenth Century: Jiao Xun, Wang Lai, and Li Rui," in Cheng-hung Lin and Daiwie Fu, eds., *Philosophy and Conceptual History of Science in Taiwan* (Dordrecht, Neth.: Kluwer Academic Publishers, 1993), pp. 191–192.
39. Bennett, *John Fryer,* p. 18, and Meng, "Hybrid Science," pp. 29–30. See also Biggerstaff, *Earliest Modern Government Schools,* p. 172.
40. Meng, "Hybrid Science," pp. 16–24, especially tables 1 and 2; and Hans van de Wen, "War in the Making of Modern China," *Modern Asian Studies* 30, no. 4 (1996): 740.
41. Meng, p. 17. See also Pong, *Shen Pao-chen,* p. 224; and Biggerstaff, *Earliest Modern Government Schools,* pp. 246–247.
42. Meng, "Hybrid Science," pp. 17–19. Kenneth Pomeranz, in his book *The Great Divergence: Europe, China, and the Making of the Modern World Economy* (Princeton, N.J.: Princeton University Press, 2001), pp. 62–68, notes that China's fossil fuels were relatively inaccessible when compared to Europe's.
43. Meng, "Hybrid Science," pp. 21–23. In 1905, the arsenal was separated into a shipyard and machine shop.
44. Pong, *Shen Pao-chen,* pp. 208–209.
45. Biggerstaff, *Earliest Modern Government Schools,* pp. 200–208. German military technology was favored after the German victory in the 1870–1871 Franco-Prussian War.
46. David Pong, *Shen Pao-chen,* pp. 241–243, 261.
47. Steven Leibo, *Transferring Technology to China: Prosper Giquel and the Self-Strengthening Movement* (Berkeley: University of California Press, 1985); Pong, *Shen Pao-chen,* pp. 214–225; and Biggerstaff, *Earliest Modern Government Schools,* pp. 203–210.
48. Biggerstaff, *Earliest Modern Government Schools,* pp. 203–211. See also Benjamin A. Elman, *From Philosophy to Philology: Intellectual and Social Aspects of Change in Late Imperial China,* 2d ed. (Los Angeles: UCLA Asia Institute Monograph Series, 2001), p. 256, and *A Cultural History of Civil*

Examinations in Late Imperial China (Berkeley: University of California Press, 2000), pp. 135, 221–222.

49. Biggerstaff, *Earliest Modern Government Schools,* pp. 214–219. See also Benjamin A. Elman, *On Their Own Terms: Science in China, 1550–1900* (Cambridge: Harvard University Press, 2005), p. 374.
50. Biggerstaff, *Earliest Modern Government Schools,* pp. 223–241. See also Kuo and Liu, "Self-Strengthening," pp. 524–525.
51. Biggerstaff, *Earliest Modern Government Schools,* pp. 53, 220, 239, 271, and Pong, *Shen Pao-chen,* pp. 208–209, 266–270. See also Kuo and Liu, "Self-Strengthening," p. 534.
52. Thomas Kennedy, *The Arms of Kiangnan: Modernization in the Chinese Ordnance Industry, 1860–1895* (Boulder, Colo.: Westview Press, 1978), pp. 150–160; and Pong, *Shen Pao-chen,* pp. 292–293, 335.
53. John Rawlinson, *China's Struggle for Naval Development, 1839–1895* (Cambridge: Harvard University Press, 1967), pp. 60–61.
54. Bruce Elleman, *Modern Chinese Warfare, 1795–1989* (New York: Routledge, 2001), pp. 82–93.
55. Rawlinson, *China's Struggle,* pp. 109–128. Compare Allen Fung, "Testing the Self-Strengthening: The Chinese Army in the Sino-Japanese War of 1894–95," *Modern Asian Studies* 30, no. 4 (1996): 1010–1015.
56. Rawlinson, *China's Struggle,* pp. 129–139, and Biggerstaff, *Earliest Modern Government Schools,* pp. 221–222.
57. Noriko Kamachi, "The Chinese in Meiji Japan: Their Interaction with the Japanese before the Sino-Japanese War," in Akira Iriye, ed., *The Chinese and the Japanese: Essays in Political and Cultural Interactions* (Princeton, N.J.: Princeton University Press, 1980), pp. 69–72. See also Donald Keene, "The Sino-Japanese War of 1894–95 and Its Cultural Effects in Japan," in Donald Shively, ed., *Tradition and Modernization in Japanese Culture* (Princeton, N.J.: Princeton University Press, 1971), pp. 122–123.
58. Foreign media accounts are presented in S. C. M. Paine, *The Sino-Japanese War of 1894–1895: Perceptions, Power and Primacy* (Cambridge, Eng.: Cambridge University Press, 2003), pp. 107–134, 138–140, 154–160. See also Marius Jansen, Samuel Chu, Shumpei Okamoto, and Bonnie Oh, "The Historiography of the Sino-Japanese War," *International History Review* 1, no. 2 (April 1979): 191–227.
59. Rawlinson, *China's Struggle,* pp. 163–169. See also Keene, "Sino-Japanese War," p. 132.
60. *Japan Weekly Mail,* August 4, 1894, pp. 130–131. See also Keene, "Sino-

Japanese War," pp. 127, 132. Compare Shumpei Okamoto, "Background of the Sino-Japanese War, 1894–95," in Okamoto, *Impressions of the Front: Woodcuts of the Sino-Japanese War, 1894–95* (Philadelphia: Museum of Art, 1983), p. 13. Compare Paine, *Sino-Japanese War,* pp. 134–140.

61. *Japan Weekly Mail,* August 4, 1894, p. 132. See also Paine, *Sino-Japanese War,* pp. 132–135, 158–163.

62. Okamoto, "Background of the Sino-Japanese War," p. 13. Compare Paine, *Sino-Japanese War,* pp. 178–192, 197–198.

63. Rawlinson, *China's Struggle,* pp. 169–174, 201. See also Biggerstaff, *Earliest Modern Government Schools,* p. 248.

64. Fung, "Testing the Self-Strengthening," pp. 1007–1031. Compare Richard Smith, "Foreign Training and China's Self-Strengthening: The Case of Feng-huang-shan," *Modern Asian Studies* 10, no. 2 (1976): 195–223.

65. Rawlinson, *China's Struggle,* pp. 174–197; and Paine, *Sino-Japanese War,* pp. 194–195, 206–222, 228–238. See also Louise Virgin, "Japan at the Dawn of the Modern Age," in *Japan at the Dawn of the Modern Age: Woodblock Prints from the Meiji Era, 1868–1912* (Boston: Museum of Fine Arts, 2001), pp. 66–72, 86.

66. Okamoto, "Background of the Sino-Japanese War," p. 16; Virgin, "Japan at the Dawn of the Modern Age," p. 112; and Paine, *Sino-Japanese War,* pp. 192–195, 247–293, 306–309. Compare J. N. Westwood, *Russia against Japan, 1904–05: A New Look at the Russo-Japanese War* (Houndmills, Eng.: Macmillan, 1986).

67. Paine, *Sino-Japanese War,* pp. 265–266, 326–327.

68. See David Wright, "John Fryer and the Shanghai Polytechnic: Making Space for Science in Nineteenth-Century China," *British Journal for the History of Science* 29 (1996): 15; and Kuo and Liu, "Self-Strengthening," p. 587. See also Zhao Hongjun, "Chinese versus Western Medicine: A History of Their Relations in the Twentieth Century," *Chinese Science* 10 (1991): 36–37.

69. Van de Ven, "War in the Making of Modern China," pp. 737–756; and Paine, *Sino-Japanese War,* pp. 64–66, 333–366. Compare Richard Smith, "Reflections on the Comparative Study of Modernization in China and Japan: Military Aspects," *Journal of the Hong Kong Branch of the Royal Asiatic Society* 16 (1976): 11–23.

70. Fryer, "Account of the Department for the Translation of Foreign Books," pp. 77–81.

71. Bennett, *John Fryer,* pp. 42–44; and Masini, *Formation of Modern Chinese Lexicon,* p. 75.

72. See Kuo and Liu, "Self-Strengthening," pp. 519–537.
73. Thomas Kennedy, "Chang Chih-tung and the Struggle for Strategic Industrialization: The Establishment of the Hanyang Arsenal, 1884–1895," *HJAS* 33 (1973): 154–182.
74. Natascha Vittinghoff, "Social Actors in the Field of New Learning in Nineteenth Century China," in Michael Lackner and Vittinghoff, eds., *Mapping Meanings: The Field of New Learning in Late Qing China* (Leiden, Neth.: E. J. Brill, 2004), pp. 75–118.
75. David Wright, "Careers in Western Science in Nineteenth-Century China: Xu Shou and Xu Jianyin," in *Journal of the Royal Asiatic Society,* 3d ser., 5 (1995): 49–80, 88.
76. Meng, "Hybrid Science," pp. 26–27. On "techno-science," see Bruno Latour, *Science in Action: How to Follow Scientists and Engineers through Society* (Cambridge: Harvard University Press, 1987), pp. 145–176.
77. Evans, "The Canton T'ung-wen Kuan," pp. 89–103.
78. On Lu Xun, see Howard Boorman and Richard Howard, eds., *Biographical Dictionary of Republican China* (New York: Columbia University Press, 1967), p. 417. See also Benjamin A. Elman, "Wang Kuo-wei and Lu Hsun: The Early Years," *Monumenta Serica* 34 (1979–1980): 389–401.
79. Biggerstaff, *Earliest Modern Government Schools,* pp. 53, 251. See also Kuo and Liu, "Self-Strengthening," p. 534.
80. Kwang-ching Liu, "Nineteenth-Century China," in Ping-ti Ho and Tang Tsou, eds., *China in Crisis,* 2 vols. (Chicago: University of Chicago Press, 1968), vol. 1, pp. 93–178.

7. The Displacement of Traditional Chinese Science and Medicine

1. Paula Harrell, *Sowing the Seeds of Change: Chinese Students, Japanese Teachers, 1895–1905* (Stanford: Stanford University Press, 1992), p. 214; and Barry Keenan, "Beyond the Rising Sun: The Shift in the Chinese Movement to Study Abroad," in Laurence Thompson, ed., *Studia Asiatica* (San Francisco: Chinese Materials Center, 1975), pp. 157–169.
2. Albert Craig, "Science and Confucianism in Tokugawa Japan," in Marius Jansen, ed., *Changing Japanese Attitudes toward Modernization* (Princeton, N.J.: Princeton University Press, 1965), pp. 139–142. See also Numata Jirō, *Western Learning: A Short History of the Study of Western Science in Early Modern Japan,* trans. R. C. J. Bachofner (Tokyo: Japan-Netherlands Institute, 1992), pp. 60–95.

3. James Reardon-Anderson, *The Study of Change: Chemistry in China, 1840–1949* (Cambridge, Eng.: Cambridge University Press, 1991), pp. 82–87.
4. See Benjamin Elman, *Classicism, Politics, and Kinship: The Ch'ang-chou School of New Text Confucianism in Late Imperial China* (Berkeley: University of California Press, 1990), p. 302.
5. Min Tu-ki, *National Polity and Local Power: The Transformation of Late Imperial China,* trans. Tim Brook and Philip Kuhn (Cambridge: Harvard University Press, 1989), p. 84.
6. James Hevia, *English Lessons: The Pedagogy of Imperialism in Nineteenth-Century China* (Durham, N.C.: Duke University Press, 2003), pp. 195–240.
7. Benjamin Elman, *A Cultural History of Civil Examinations in Late Imperial China* (Berkeley: University of California Press, 2000), pp. 608–618.
8. Kung-chuan Hsiao, *A Modern China and a New World: Kang Yu-wei, Reformer and Utopian, 1858–1927* (Seattle: University of Washington Press, 1975), pp. 328–346; and Douglas Reynolds, *China, 1898–1912: The Xinzheng Revolution and Japan* (Cambridge: Harvard University Press, 1993), pp. 43–44.
9. Elman, *A Cultural History,* pp. 585–594, 608–618.
10. Hsiao, *A Modern China,* pp. 306–307, 331–346. See also *ECCP,* pp. 703–704.
11. Mingjie Hu, "Merging Chinese and Western Mathematics: The Introduction of Algebra and the Calculus in China, 1859–1903," Ph.D. diss., Princeton University, 1998, pp. 271–279, 282–283.
12. David Wright, "Tan Sitong and the Ether Reconsidered," *Bulletin of the School of Oriental and African Studies* 57 (1994): 551–557. Compare Daniel Siegel, "Thomson, Maxwell, and the Universal Ether," in G. N. Cantor and M. J. S. Hodge, eds., *Conceptions of Ether: Studies in the History of Ether Theories, 1740–1900* (Cambridge, Eng.: Cambridge University Press, 1981), pp. 245–246.
13. Stewart followed up with a sequel to his *Unseen Universe,* 9th ed. (London: Macmillan, 1880) called *Paradoxical Philosophy* (London: Macmillan, 1878).
14. Hu, "Merging Chinese and Western Mathematics," pp. 232–285. See also Knight Biggerstaff, *The Earliest Modern Government Schools in China* (Ithaca: Cornell University Press, 1961), p. 171.
15. Tian Miao, "The Westernization of Chinese Mathematics: A Case Study of the *duoji* Method and Its Development," *East Asian Science, Technology, and Medicine* 20 (2003): 63–70.
16. Paul Bailey, *Reform the People: Changing Attitudes towards Popular Educa-*

tion in Early Twentieth Century China (Edinburgh: Edinburgh University Press, 1990), pp. 140–141.

17. Hu, "Merging Chinese and Western Mathematics," pp. 349–364. See also Marianne Bastid, *Educational Reform in Early Twentieth-Century China,* trans. Paul J. Bailey (Ann Arbor: University of Michigan China Center, 1988), pp. 38–39.
18. Bastid, *Educational Reform,* pp. 12–13.
19. Zuoyue Wang, "Saving China through Science: The Science Society of China, Scientific Nationalism, and Civil Society in Republican China," *Osiris* 33 (2002): 291.
20. Paul Unschuld, "Epistemological Issues and Changing Legitimation: Traditional Chinese Medicine in the Twentieth Century," in Charles Leslie and Allan Young, eds., *Paths to Asian Medical Knowledge* (Berkeley: University of California Press, 1992, pp. 47–49; and Zhao Hongjun, "Chinese versus Western Medicine: A History of Their Relations in the Twentieth Century," *Chinese Science* 10 (1991): 24, 26.
21. Yi-li Wu, "Transmitted Secrets: The Doctors of the Lower Yangzi Region and Popular Gynecology in Late Imperial China," Ph.D. diss., Yale University, 1998.
22. Angela Leung, "Organized Medicine in Ming-Ch'ing China," *Late Imperial China* 8, no. 1 (1987): 134–166. Compare Kerrie MacPherson, *A Wilderness of Marshes: The Origins of Public Health in Shanghai, 1843–1893* (Oxford: Lexington Books, 2002), pp. 49–82.
23. Carl F. Nathan, "The Acceptance of Western Medicine in Early Twentieth Century China: The Story of the North Manchurian Plague Prevention Service," in John Bowers and Elizabeth Purcell, eds., *Medicine and Society in China* (New York: Josiah Macy Jr. Foundation, 1974), pp. 55–75. See also Carol Benedict, "Policing the Sick: Plague and the Origins of State Medicine in Late Imperial China," *Late Imperial China* 14, no. 2 (December 1993): 60–77.
24. Zhao, "Chinese versus Western Medicine," pp. 27–28.
25. Ralph Crozier, *Traditional Medicine in Modern China: Science, Nationalism, and the Tensions of Cultural Change* (Cambridge: Harvard University Press, 1968), pp. 151–209. See also Bridie Andrews, "Traditional Chinese Medicine as Invented Tradition," *Bulletin of the British Association for Chinese Studies* 6 (1995): 6–15.
26. Zhao, "Chinese versus Western Medicine," pp. 33–34.
27. D. C. Epler, "Bloodletting in Early Chinese Medicine and Its Relation to

the Origin of Acupuncture," *Bulletin of the History of Medicine* 54 (1980): 337–367.

28. Bridie Andrews, "Tailoring Tradition: The Impact of Modern Medicine on Traditional Chinese Medicine, 1887–1937," in Viviane Alleton and Alexeï Volkov, eds., *Notions et Perceptions du Changement en Chine* (Paris: Collège de France, Institut des hautes études chinoises, 1994), pp. 154–155, 158–159. For other fields, see Laurence Schneider, "Genetics in Republican China," in John Bowers, J. William Hess, and Nathan Sivin, eds., *Science and Medicine in Twentieth-Century China: Research and Education* (Ann Arbor: University of Michigan Center for Chinese Studies, 1988), pp. 3–29; and Yang Tsui-hua, "The Development of Geology in Republican China, 1912–1937," in Cheng-hung Lin and Daiwie Fu, eds., *Philosophy and Conceptual History of Science in Taiwan* (Dordrecht, Neth.: Kluwer Academic, 1993), pp. 221–244.

29. Bridie Andrews, "Tuberculosis and the Assimilation of Germ Theory in China, 1895–1937," *Journal of the History of Medicine and Allied Sciences* 52, no. 1 (1997): 114–157; and Hugh Shapiro, "The Puzzle of Spermatorrhea in Republican China," *positions* 6, no. 3 (Winter 1998): 551–596.

30. Federico Masini, *The Formation of Modern Chinese Lexicon and Its Evolution toward a National Language: The Period from 1840 to 1898* (Berkeley: University of California, *Journal of Chinese Linguistics* Monograph Series, 1993), pp. 84–88.

31. Annick Horiuchi, *Les Mathématiques japonaises, à l'époque d'Edo* (Paris: Vrin, 1994), pp. 119–155.

32. Tatsuhiko Kobayashi, "What Kind of Mathematics and Terminology Was Transmitted into Eighteenth-Century Japan from China?" *Historia Scientiarum* 12, no. 2 (2002): 1–17.

33. Kume Kunitake, comp., *The Iwakura Embassy 1871–73: A True Account of the Ambassador Extraordinary and Plenipotentiary's Journey of Observation through the United States of America and Europe,* 5 vols. (Matsudo: The Japan Documents, 2002), vol. 5, p. 352.

34. David Wright, "The Translation of Modern Western Science in Nineteenth-Century China, 1840–1895," *Isis* 89 (1998): 671; and Masini, *Formation of Modern Chinese Lexicon,* pp. 91–92. See also Takehiko Hashimoto, "Introducing a French Technological System: The Origin and Early History of the Yokosuka Dockyard," *East Asian Science, Technology, and Medicine* 16 (1999): 53–65.

35. Tsuen-hsuin Tsien, "Western Impact on China through Translation," *Far Eastern Quarterly* 13 (1954): 323–325.

36. Reynolds, *China,* pp. 48, 58–61; and Keenan, "Beyond the Rising Sun," p. 157.
37. Masini, *Formation of Modern Chinese Lexicon,* pp. 145–151.
38. Timothy Weston, *The Position of Power: Beijing University, Intellectuals, and Chinese Culture, 1898–1929* (Berkeley: University of California Press, 2004), pp. 50–52.
39. Iwo Amelung, "Naming Physics: The Strife to Delineate a Field of Modern Science in Late Imperial China," in Michael Lackner and Natascha Vittinghoff, *Mapping Meanings: The Field of New Learning in Late Qing China* (Leiden, Neth.: E. J. Brill, 2004), pp. 381–422.
40. Compare Reynolds, *China,* pp. 65–110, 131–150.
41. Wang Bing, "On the Physics Terminology in Chinese and Japanese during Early Modern Times," in Yung Sik Kim and Francesca Bray, eds., *Current Perspectives in the History of Science in East Asia* (Seoul: Seoul National University Press, 1999), pp. 517–521.
42. Compare Hildred Geertz, "An Anthropology of Religion and Magic, I," *Journal of Interdisciplinary History* 6, no. 1 (Summer 1975): 71–89.
43. Weston, *Position of Power,* p. 83.
44. Michael Adas, "Contested Hegemony: The Great War and the Afro-Asian Assault on the Civilizing Mission Ideology," *Journal of World History* 15, no. 1 (March 2004): 31–63.
45. Chow Tse-tsung, *The May 4th Movement: Intellectual Revolution in Modern China* (Cambridge: Harvard University Press, 1960), pp. 327–329.
46. In *The Way and the Word: Science and Medicine in Early China and Greece* (New Haven: Yale University Press, 2002), authors Geoffrey Lloyd and Nathan Sivin provide more balanced comparisons.
47. Toby E. Huff, *The Rise of Early Modern Science: Islam, China, and the West,* rev. ed. (Cambridge, Eng.: Cambridge University Press, 2003). See my review of the 1993 edition in *American Journal of Sociology* (November 1994): 817–819.

Appendix 1

The "Ten Computational Classics" were used at the Imperial Academy to teach mathematics during the Sui and Tang dynasties. See Siu Man-Keung and Alexeï Volkov, "Official Curriculum in Traditional Chinese Mathematics: How Did Candidates Pass the Examinations?" *Historia Scientiarum* 9, no. 1 (1999): 85–99.

1. See Christopher Cullen, *Astronomy and Mathematics in Ancient China: The Zhoubi suan jing* (Cambridge, Eng.: Cambridge University Press, 1996), pp. 171–205; and Li Yan and Du Shiran, *Chinese Mathematics: A Concise History,* trans. John Crossley and Anthony Lun (Oxford: Clarendon Press, 1987), pp. 25–32.
2. See Shen Kangshen, John N. Crossley, and Anthony Lun, trans., *The Nine Chapters on the Mathematical Art: Companion and Commentary* (Oxford: Oxford University Press; Beijing: Science Press, 1999); and Karine Chemla and Guo Shuchun, trans., *Les Neuf Chapitres: Le Classique Mathématique de la Chine ancienne et ses commentaires* (Paris: Dunod, 2004). See also Donald B. Wagner, "An Early Derivation of the Volume of a Pyramid: Liu Hui, Third Century A.D.," *Historia Mathematica* 6 (1979): 164–188.
3. Frank Swetz, trans., *The Sea Island Mathematical Manual: Surveying and Mathematics in Ancient China* (University Park: Pennsylvania State University Press, 1992). See also Alexander Wylie, *Notes on Chinese Literature* (Shanghai: American Presbyterian Mission Press, 1867), p. 114; and Li Yan and Du Shiran, *Chinese Mathematics: A Concise History,* trans. John Crossley and Anthony Lun (Oxford: Clarendon Press, 1987), pp. 75–80.
4. Wylie, *Notes on Chinese Literature,* p. 114; and Jean-Claude Martzloff, *A History of Chinese Mathematics,* trans. Stephen Wilson (New York: SpringerVerlag, 1997), pp. 136–138. See also Li and Du, *Chinese Mathematics,* pp. 92–95.
5. Li and Du, *Chinese Mathematics,* pp. 95–97; Wylie, *Notes on Chinese Literature,* p. 114; and Martzloff, *History of Chinese Mathematics,* p. 124.
6. Li and Du, *Chinese Mathematics,* pp. 97–98, Wylie, *Notes on Chinese Literature,* p. 115; and Martzloff, *History of Chinese Mathematics,* p. 141.
7. Li and Du, *Chinese Mathematics,* pp. 98–100, Wylie, *Notes on Chinese Literature,* p. 115, and Martzloff, *History of Chinese Mathematics,* pp. 124, 138–139.
8. Wylie, *Notes on Chinese Literature,* p. 115; and Martzloff, *History of Chinese Mathematics,* pp. 124, 140.
9. Li and Du, *Chinese Mathematics,* pp. 100–104; Wylie, *Notes on Chinese Literature,* p. 115–116; and Martzloff, *History of Chinese Mathematics,* pp. 125, 140–141.

Appendix 3

1. On the geography texts, see the discussion in Zou Zhenhuan, *Wan Qing Xifang dilixue zai Zhongguo* (Western geography in China during the late Qing) (Shanghai: Guji chubanshe, 2000), pp. 126–133.

Appendix 5

1. The translation for this question is taken from John Fryer, "Chinese Prize Essays: Report of the Chinese Prize Essay Scheme in Connection with the Chinese Polytechnic Institution and Reading Rooms, Shanghai, for 1886 and 1887," *North-China Herald,* January 25, 1888, p. 100. Fryer's "Ching-kang-ching" is the Later Han classicist Zheng Kangcheng, i.e., Zheng Xuan (127–200), whose classical importance Fryer did not seem to recognize.

ACKNOWLEDGMENTS

Margaret Jacob of UCLA encouraged me to think of my larger study of science in late imperial China as a first step in preparing a more general textbook for the Themes in the History of Science, Medicine, and Technology series published by Harvard University Press. At the Press, Ann Downer-Hazell, editor in Life Sciences and Health, kept me focused on both projects. And Julie Carlson, my manuscript editor, helped make the textbook clearer for readers. I also want to thank Robert Mowry from the Sackler Museum at Harvard for his remarkable Chinese Ceramics Workshop at Princeton University on May 25–27, 2005.

INDEX